# A T
# TROV
# FOR TV FANS!

***TVacations*** tells you:

- Why a trip to New York's Greenwich Village is like a walk through the backlot of a Hollywood studio: *Mad About You*, *Dream On*, and *The Cosby Show* all filmed their residential exteriors within walking distance, as well as the garage for *Taxi*, and the 9th Precinct station used for both *Kojak* and *NYPD Blue!*

- How television influences decisions in the Oval Office: *Bonanza* was so popular that President Johnson would *not* schedule speeches on Sunday at 9 P.M. because he reportedly thought that too many viewers (and voters) would be offended!

- Where our youth really get their education: In a 1991 documentary about Kent State, every student interviewed knew the words to *The Brady Bunch* theme!

- Why you can fly to the Nantucket Memorial Airport (from *Wings*) and journey to Vermont's Waybury Inn (from *Newhart*), but *can't* visit Angela Lansbury's *Murder, She Wrote* New England coastal hometown: It's in California!

- How life imitates art: the residents of Mount Airy, North Carolina, were so convinced that their hamlet was the basis for *The Andy Griffith Show*'s Mayberry that they actually erected Floyd's City Barber Shop, The Mayberry Bed & Breakfast, The Mayberry Motor Inn, and Aunt Bea's Barbecue!

This is only a taste of the delights awaiting television devotees in

## TVacations

# TVacations

## A Fun Guide to the Sites, the Stars, and the Inside Stories Behind Your Favorite TV Shows

**Fran Wenograd Golden**

POCKET BOOKS
New York London Toronto Sydney Tokyo Singapore

*To my three E's,*
*Eddie, Erin and Eli*



An *Original* Publication of POCKET BOOKS

POCKET BOOKS, a division of Simon & Schuster Inc.
1230 Avenue of the Americas, New York, NY 10020



Golden, Fran Wenograd.
TVacations : a fun guide to the sites, the stars, and the inside stories behind your favorite TV shows / Fran Wenograd Golden.
p. cm.
ISBN 0-671-89024-7
1. Television program locations—United States—Guidebooks.
2. Television programs—United States—Miscellanea. I. Title.
PN1992.78.A1G66 1996
384.55'4'02573 dc20 95-42987
CIP

First Pocket Books trade paperback printing April 1996

10 9 8 7 6 5 4 3 2 1



Cover design by Patrice Kaplan; cover illustration by Cindy Sandro; photos from *I Love Lucy, The Cosby Show* and *Seinfeld* courtesy of Photofest; photos from *Happy Days* and *I Dream of Jeannie* courtesy of Motion Picture and TV Photo Archives

Text design by Stanley S. Drate/Folio Graphics Co. Inc.

Printed in the U.S.A.

# Acknowledgments

The author owes a big thank-you to:

the many people around the country who contributed to making this book possible including state and city film commission officials, local and state tourism officials, hotel employees, museum directors, local police personnel and other town and city employees, librarians, real estate agents, representatives of the TV networks, historians, and federal employees. Their prompt and friendly responses to my requests for information were greatly appreciated.

the staff at the Museum of Television and Radio in New York, especially researcher Jonathan Rosenthal, whose fax responses to my fax requests were a great help.

those who provided photos and who showed me the sights, including Ben and Jen, Eric L., Eric T., and James.

my wonderful bosses at my day job, Nadine Godwin and Alan Fredericks, who offered their support and encouragement. And the many among my co-workers at *Travel Weekly* who offered their expertise and assistance to this project.

my editors at Pocket Books, especially Eric Tobias, for seeing me through the early stages of this, my first book, and for buying the book in the first place. And Amy Einhorn for her thoughtful editing.

Jeremy Solomon, my agent, for his encouragement, persistence, and insight.

Kate Patterson for her research assistance and for making my family tasks easier during the final stages of this project.

my friends and relatives for putting up with my terse responses to their phone calls during the final stages of writing this book.

my husband and kids for their patience with this project, and for the joy they add to my life on a daily basis.

and last but not least, my mom, Marlene, for her support and free baby-sitting, and especially for telling me she's proud of me.

# CONTENTS

## NEW YORK CITY ENVIRONS



## BALTIMORE, PHILADELPHIA, AND WASHINGTON, D.C.



## THE SOUTH

## FLORIDA



## THE MIDWEST



## CHICAGO



## THE WEST AND THE OLD WEST

## ARIZONA AND NEVADA



## LOS ANGELES



## LOS ANGELES COUNTY AND ENVIRONS

# INTRODUCTION

Ever wonder what it would be like to sit in Norm's chair at the *Cheers* bar and toss back a couple of beers, or order a round for the gang. Where are the pretzels, Sam? Or strut into a bar in Dallas with a ten-gallon hat like J.R. on *Dallas*, and order a "bourbon and branch"?

Well, here's your chance.

It's time for us television gazers to get off the couch and visit those familiar television scenes we know and love.

Take a walk on the dark side in Snoqualmie, Washington, home of *Twin Peaks*. You can check into the Great Northern Hotel (in real life, the Salish Lodge), and maybe even have a chat with a log lady.

Or how about stopping in at the luxurious estate seen as the home of the wealthy Carrington family on *Dynasty* (it's actually a museum in Woodside, California) and see why the house inspired Alexis's villainous jealousy.

Or pay a nostalgic tribute to *Laverne & Shirley* by doing a brewery tour in Milwaukee. And don't forget to stop by a bowling alley for a slice and a Coke.

Check out the Cartwright ranch in Incline Village, Nevada, and experience the Old West seen on *Bonanza*. Or take in quaint Boulder, Colorado, like only an Ork-an, as on *Mork & Mindy*, could.

From the back woods to the big cities, *TVacations* is your guide to the real-life settings of your favorite shows from the little screen.

We've researched the actual locations seen on selected shows of the past thirty years, including the addresses seen in exterior shots, which are used to help establish the setting for the action on the shows.

A lot of the information in this book is the kind of stuff that is only known by the show's production crew and local residents at each location. And now you can use this book to find the actual sites and sounds across the United States.*

You can view the exterior of the real house seen as the Huxtables' on *The Cosby Show,* see the police station that's featured on *Kojak* and *NYPD Blue* (it's the same one), and see the outside of the funky Victorian house where Mary Richards rents an apartment in Minneapolis on *The Mary Tyler Moore Show.*

In cases where real addresses are not available, we offer suggestions that will help you pay tribute to the memories of your favorite fictional television characters.

You can take a walking tour of Mount Airy, North Carolina, for instance. The town is Andy Griffith's boyhood home and believed to have inspired fictional Mayberry on *The Andy Griffith Show* (a lot of the names in town are the same as on the show, as we've noted).

Or visit Riverside, Iowa, which bills itself as the future hometown of Captain Kirk on *Star Trek,* and has a replica of the starship *Enterprise* on display.

Although times of operation are listed for locations that are open to tourists, such as parks, museums, hotels, restaurants, and retail businesses, it's always a good idea to call ahead in case of any changes. If admission is charged, we've noted it. Where applicable, we've also made suggestions on where to buy souvenirs for the shows. And we've included our favorite trivia tidbits for your enjoyment.

You can read this book for fun, and beat all your pals in the television category of Trivial Pursuit.

Or, better yet, you can get off the couch and see the sites yourself. But remember, don't leave your TV friends far behind.

---

*In cases where the locations are private houses, or businesses that do not cater to tourists, visitor discretion is advised. You can take pictures on the outside, but please don't bother those inside.

# BOSTON

This city may be known for its Irish-American politicians, sports heroes, and institutions of higher education, but to most television viewers, its place in history belongs to *Cheers*.

Few visitors miss a chance to see the real-life pub shown in the opening credits of the show, the Bull & Finch Pub. And some diehard fans have even commented that they think that coming to Boston is coming to *Cheers*.

The pub is, in fact, the second most popular tourist attraction in Boston, after Old Ironsides (the U.S.S. *Constitution*), beating out historic sites on the famous Freedom Trail.

Bull & Finch owner Tom Kershaw, who has parlayed the show's popularity into a multimillion-dollar corporation selling *Cheers* merchandise, considered changing the name of his pub to *Cheers*. But he caved in to local resistance and today simply hangs a *Cheers* flag outside his establishment.

While *Cheers* may be number one here, the city of Boston is not without other sights of interest to TV fans, including the building seen as St. Eligius Hospital on *St. Elsewhere*, and the various haunts of private detective Spenser on *Spenser: For Hire*.

## CHEERS
**(September 1982–May 1993, NBC)**

Once upon a time, producers Jim Burrows and Glen and Les Charles came to Boston with an idea for a new comedy series and in search of inspiration.

In the American hometown of Irish pubs and student hangouts the team that had earlier worked together on *Taxi* looked for an all-American neighborhood bar—the kind of place where talk tends to politics and sports and "where everybody knows your name."

Visiting about half a dozen local bars, the producers found they were most attracted to the Bull & Finch Pub, a basement bar off the Public Garden.

The name itself was part of the draw, as was what they found inside, including longtime bartender Eddie Doyle. Some believe Doyle was the model for "Coach," the original *Cheers* bartender, played by Nicholas Colasanto until his death in 1985. Like Coach, Doyle is known to be good-hearted, and to have a great sense of humor.

Shots of the real pub's exterior were used on the show—the producers simply added "Cheers" and "Melville's Fine Seafoods" signs. And while the series was actually shot in Hollywood, devoted fans flock to the Bull & Finch as the show's true home.

The show featured very witty repartee between the main characters, who spent a great deal of their time at the bar.

The show also featured its Boston setting. The producers added local color with guest shots on *Cheers* by such Boston political luminaries as former U.S. House Speaker "Tip" O'Neill (his former secretary's son was a casting director for the show), Massachusetts governor (and later presidential candidate) Michael Dukakis, and Boston mayor (and later U.S. envoy to the Vatican) Ray Flynn.

Boston sports heroes also stopped by *Cheers*, including Boston Celtic Kevin McHale (the *Cheers* gang had him sneak them into the Boston Garden to count how many bolts hold down the parquet floor) and baseball's top hitter, Wade Boggs.

Fans of the series will notice on entering the Bull & Finch that it's not the same familiar spot inside as depicted on the show's set. With crowds of tourists frequenting the Bull & Finch, many of the bar's real original local followers have moved on to other lesser known local watering holes, taking away a bit of the local color.

But that doesn't mean cries of "Norm!" can't occasionally be heard.

The *Cheers* bar in Boston. *(Photo by Photography Unlimited, courtesy of the Hampshire House.)*

## WHAT TO SEE IN BOSTON, MASSACHUSETTS

▶ **The Bull & Finch Pub,** 84 Beacon Street (across from the Public Garden). Once used as a flower-arranging room in the home of a wealthy Boston family, the pub is a bit cramped, and *Cheers* fans will notice immediately that the mahogany bar is not in the

Inside the Bull & Finch Pub. *(Photo by Photography Unlimited, courtesy of the Hampshire House.)*

center, as on the show, but against a wall. The real-life back room is not used for pool, but rather for seating guests. As a spoof, however, the proprietor has placed a little plastic pool table on the ceiling. There's a manager's office, but it is down a corridor, not off the bar, and is not open to the public (nor is it like Sam's office, we're told). Drinks and pub food are served here daily, from 11 A.M. to 2 A.M. Call 617-227-9605.

▶ **Hampshire House/Library Grill,** 84 Beacon Street (upstairs from the Bull & Finch). Yes, there is a classier restaurant upstairs from the bar, and while it's not called Melville's, as on the show, seafood is a specialty here. Done up in turn-of-the-century decor, the restaurant is open for dinner seven days a week, 5:30 P.M. to 10:30 P.M. Live music is offered six days a week for dancing on a small dance floor. The restaurant also offers a Sunday Jazz Brunch from 10:30 A.M. to 2:30 P.M. On the first floor are framed photos from the show and a *Cheers* souvenir store, with T-shirts, mugs, and other items. Call 617-227-9600.

▶ **Cheers Souvenir Shops** are located throughout Boston including stores at Faneuil Hall and Logan International Airport. *Cheers* items are also available by mail order. For a catalog, call Bull & Finch Enterprises, 800-962-3333, or send a fax, 508-657-5760.

▶ **Fenway Park,** 4 Yawkey Way. The ballpark may not feature Sam "Mayday" Malone on the mound, but sports fans will want to check out this historical ballpark. For Boston Red Sox tickets call 617-267-1700.

## Fun Facts

• Jay Leno did a rare remote broadcast of *The Tonight Show* from the Bull & Finch on the final night of *Cheers* in 1993, but our sources tell us it wasn't his first visit to the bar. He visited the Bull & Finch while a student at nearby Emerson College.

• Norm's (George Wendt) real name on the show was Hilary (the name of Wendt's real-life daughter). Norm explained in one episode that he's named after his grandfather.

• Cliff's (John Ratzenberger) know-it all lines included such observations as "Due to the shape of the North American elk's esophagus, even if it could speak, it could not pronounce the word *lasagna*."

• While Norm's wife Vera's face was never seen on the show, her voice was that of Wendt's real wife, Bernadette Birkett.

• Yes, Sam "Mayday" Malone (Ted Danson) wore a hairpiece, it is revealed in a final episode. So does Danson, in case you didn't get it.

• The beer pouring out of the taps on the show was non-alcoholic, not the real thing. For the last six years of the show the brand used was Kingsbury.

• Among Carla's (Rhea Perlman) eleven kids were twins named after Elvis Presley and "his dead twin brother Jesse."

• Frasier (Kelsey Grammer) went on to his own hit spin-off show, *Frasier*. His wife on *Cheers*, Lilith (Bebe Neuwirth) guest-starred on *Frasier*.

• Sam (Ted Danson) was a lady killer and recovering alcoholic, but in real life, the actor was known as being rather saintly by Hollywood standards—a devoted husband and father, with no publicized addictions. That is until 1992, when he went off to work on the movie *Made in America* and fell in love with costar Whoopi Goldberg, and left his wife.

• Woody Harrelson, who played spacey bartender Woody on the show, had the womanizer reputation in real life, which he did nothing to dispute. He once told *Playboy* magazine, "If a woman says, 'Have you slept with a lot of women?' I'll say yes. If she asks, 'Are we going to see each other again?' I'll say I don't know."

• Both Actress Kirstie Alley, who played Rebecca, and Kelsey Grammer, who played Frasier, had past substance abuse problems, she with cocaine and he with drugs and alcohol. Grammer, who studied acting at Julliard, served thirty days in prison in 1990 on drug-related charges.

• Actress Bebe Neuwirth, who played Lilith, once told an interviewer that as a child she wanted to grow up to be Morticia Addams *(The Addams Family)*.

• As of 1993, sales by Bull & Finch Enterprises, which markets *Cheers* merchandise, had included 175,000 *Cheers* hats, 450,000 sweatshirts, 675,000 glassware items, and 2,000,000 T-shirts.

• Making guest appearances on *Cheers* were Tom Skerritt *(Picket Fences)*, as Evan Drake, and Jay Thomas *(Love & War)* as Carla's husband, Eddie LeBec.

• Actress Rhea Perlman, who played Carla, is married in real life to Danny DeVito *(Taxi)*.

• Frasier (Kelsey Grammer) delivered one of the best bits on the show, when he stole a line from a popular disco tune and declared "Everybody have fun tonight, everybody wang chung tonight."

• Fans were torn between loyalties to intellectual Diane, played by Shelley Long, who left the show in 1987 to pursue a film career, and her successor, brassy Rebecca (Kirstie Alley). Alley actually appeared on more shows, 154, compared to Long's 124. Both actresses appeared in the show's finale.

## ST. ELSEWHERE
**(October 1982–May 1988, NBC)**

"St. Elsewhere" is a real buzzword used by doctors when they can't remember where a referral came from. But to television viewers it is St. Eligius, the Boston teaching hospital that takes in all strays, where staff and interns work under horrible conditions and where patients have very little chance of surviving.

During the show's six-year run, faithful fans of *St. Elsewhere* saw penetrating social satire, with the show taking on such delicate topics as lovemaking on an autopsy slab, the afterlife, politics, AIDS, religion, drug abuse, and both homosexual and heterosexual rape.

*St. Elsewhere* was hailed by critics as a groundbreaker, and by some as the best television drama ever.

In researching *St. Elsewhere,* cocreators John Falsey and Joshua Brand *(Northern Exposure)* visited real hospitals including the Cleveland Clinic, St. Vincent's Hospital in New York, and the County Medical Center in Los Angeles.

Falsey and Brand were quoted as saying they wanted *St. Elsewhere* to show realism, and believed real hospitals didn't feel like those in earlier medical shows such as *Marcus Welby, M.D.* or *Medical Center.*

"It's bloody rages, music in the operating room, Bruce Springsteen blasting out of the speakers," Falsey told Associated Press.

St. Eligius Hospital was not the healthiest place to be for patients. "If you checked into St. Eligius, you died. If you were healthy and you accidently got checked in, we killed you," series executive producer Bruce Paltrow was once quoted as saying.

Even Dr. Donald Westphall (Ed Flanders), the hospital's saintly chief of staff, was not immune to the show's dark comedy. He displayed his feelings about the hospital's new management in one episode by dropping his trousers and mooning his new boss, and left with the memorable line, "You can kiss my ass."

In the finale of the series, it was suggested that the hospital and all inside, existed only in the mind of Tommy, Dr. Westphall's autistic son, who was seen looking at the St. Eligius building in a snow globe toy.

Producer Paltrow called the ending "excitingly existential."

## WHAT TO SEE IN BOSTON, MASSACHUSETTS

▶ **Franklin Square House,** 11 East Newton Street. Seen as St. Eligius Hospital on the show, this building was described by one architectural writer as "big and blunt." Built in 1868, the factory-sized building is of French Second Empire design. Originally the St. James Hotel, which housed such prestigious guests as Grover Cleveland, and later home of the New England Conservatory of Music, the building is now used for elderly housing. It's never been a hospital.

▶ **Boston City Hospital,** 818 Harrison Ave. While the Franklin Square House structure is impressive for its familiarity, fans looking for a better replication of the real-life hospital drama at St. Eligius may want to visit Boston City Hospital, the closest thing to a St. Elsewhere in Boston.

## FUN FACTS

• Dr. Mark Craig (William Daniels), the crotchety chief of surgery, who was worshiped by Dr. Victor Ehrlich (Ed Begley, Jr.), had such wonderful bullying lines as (to a female surgeon) "Why don't you go home and do the wash!"

• Actor William Daniels's real-life wife, Bonnie Bartlett, played Craig's wife, Ellen.

• The show paid homage to its producer, MTM Enterprises, in one episode, when a male patient who doesn't remember his name watches *The Mary Tyler Moore Show* and gets a flash. He runs into the corridor shouting, "I know who I am! I'm Mary Richards."

• Dr. Morton Chegley, Julia Baker's boss on *Julia*, must have headed east from L.A. for a new job. He was often paged on the St. Eligius public address system.

• Paranoid Elliott Carlin (Jack Riley) of *The Bob Newhart Show,* evidently wasn't cured by Psychologist Bob. He spent time in the psychiatric ward at St. Eligius.

• Among the greatest plot lines on the show was the death of whiny Mrs. Hufnagel (Florence Halop) when her hospital bed simply folded up on her.

• In one episode, a severed head was inexplicably mailed to Dr. Craig's mother-in-law. She dropped dead of shock after opening the package.

• Womanizer Dr. Robert Caldwell (Mark Harmon) got his face slashed by a crazy blonde with a razor blade and later contracted AIDS and died. In real life, Actor Harmon married Pam Dawber, Mindy on *Mork & Mindy.*

• Actor Denzel Washington, who played Dr. Phillip Chandler, later became an Oscar-winning movie star.

• Howie Mandel, who played Dr. Wayne Fiscus (who once died on the show but came back to life), is a stand-up comic, whose routine has included wearing surgical gloves on his head.

• Playing the wife of kindly Dr. Jack Morrison on the show was Patricia Wettig *(thirtysomething).* After his wife left him, Morrison's girlfriend was played by Helen Hunt *(Mad About You).*

## SPENSER: FOR HIRE
**(September 1985–September 1988, ABC)**

This detective show featured actor Robert Urich as Spenser, a literate and highly principled, ex-boxer, gourmet cook of a detective.

The character was created by writer Robert Parker, and has appeared in more than two dozen books. And while the show doesn't duplicate Parker's character exactly, both the print and TV Spenser make their home in Boston.

*Spenser: For Hire* was shot on location in Beantown. It was produced by John Wilder, who had earlier worked on *The Streets of San Francisco,* another show shot on location. *Spenser: For Hire* appeared at a time when shooting on location was hot, given the success of the cop show *Miami Vice.*

Parker, who himself is known as a gourmet cook, sold rights to his characters to the TV producers, and then backed off from the productions. In an interview, Parker said he almost never saw the show, although he did read the scripts.

"I wasn't a great fan," Parker said. "I knew when I took the money they wouldn't improve things, and they didn't."

*Spenser: For Hire* drew fans from in and around Boston, but it was less successful elsewhere. The show was saved from cancellation after its second year, only after an outcry of Boston-area public support, including a visit to the set by Massachusetts governor Michael Dukakis.

Studio shots for the show were filmed in a special studio set up in the Charlestown Navy Yard (near the U.S.S. *Constitution*). And many Boston landmarks appeared on the series including the Boston Common, the Combat Zone (red-light district), and the Hatch Shell on the Esplanade.

One of the show's best chase scenes took place under the now dismantled elevated trains that ran in the South End on Tremont Street. And another chase scene was filmed in boats on the Charles River.

Parker and his wife, Joan, later decided to work on their own TV adaptations of the Spenser character. They cowrote scripts for TV movies based on Parker's books *Ceremony* and *Pale Kings and Princes* in 1993, and *The Judas Ghost* and *A Savage Place* in 1994. All aired on the Lifetime channel. To save costs, Parker said, the movies were shot in Toronto, not Boston.

### What to See in Boston, Massachusetts

▶ **Mount Vernon Avenue near Charles Street, Beacon Hill.** In the pilot for the series, Spenser's South End house is burned, but he saves the life of a fireman and is rewarded by the Boston Fire Department with the use of an old firehouse on Beacon Hill. This building, shown as Spenser's abode, is actually a historic fire station, but it's now used for storage, not housing.

▶ **Ristorante Toscano's,** 41–47 Charles Street. Shown in the opening shots of the show, this is the place Spenser liked to take Susan for a good meal and intellectual discussion. The production crew for the show also liked to hang out here. The restaurant is open for lunch every day but Sunday, 11:30 A.M. to 2:30 P.M., and daily for dinner, from 5:30 P.M. to 10 P.M. Call 617-723-4090.

### What to See in Charlestown, Massachusetts

▶ **Monument and Main streets.** On the show, the city eventually needed the firehouse back so Spenser moved to this historic apartment building located above some stores. In real life, the move was precipitated by Beacon Hill residents who tired of location shooting in their neighborhood. The real building's downstairs offerings include Sullivan's Pub, a neighborhood bar.

### Fun Facts

• Spenser's powerful yet sensitive, Magnum-carrying sidekick, Hawk, played by Avery Brooks, went on to his own spin-off show, *A Man Called Hawk,* which was based in Hawk's hometown of Washington, D.C.

• Susan Silverman (Barbara Stock), Spenser's therapist girlfriend, left for a season but returned for the show's final year. During her absence, Spenser dated Assistant D.A. Rita Fiori (Carolyn McCormick).

• In the Spenser books, the detective lives on the first block of Marlborough Street, down from the Public Garden. In the most recent books, Spenser's office, which has moved around a bit, is located at the corner of Boylston and Berkeley streets, above a bank.

• The town of Spencer, Massachusetts, changed the spelling of its name from *Spencer* to *Spenser* for a day in honor of the TV detective with no first name.

# NEW ENGLAND

*Newhart,* with its Vermont country inn setting; *Murder, She Wrote,* with its scenic Maine seaside landscape; and *Wings,* with its quaint Massachusetts island backdrop, represent the New England that TV viewers probably expect to see.

But surprise! In TV-land, not so long ago New England was also home to vampires *(Dark Shadows),* very naughty small town folk *(Peyton Place),* ghosts *(The Ghost and Mrs. Muir),* and witches *(Bewitched).* And the region is supposed to be so charming and sedate!

The Vermont inn seen on *Newhart* is very real and just as lovely as on the show, and assorted islanders can be found at the real island airport in Nantucket, Massachusetts, as seen on *Wings.* Interestingly some of the eery sites on *Dark Shadows* are also real, and are located in opulent Newport, Rhode Island. But most of the settings featuring New England are either make-believe or not in New England.

*Murder, She Wrote* captured the New England seaside beauty of fictional Cabot Cove, Maine, by filming in Northern California. That's really the Pacific Ocean viewers see!

*Peyton Place* took its inspiration from a real town in New Hampshire, but was filmed on a Hollywood set. So was *I Love Lucy,* although in the show's plot, wacky Lucy and Ricky move to posh Westport, Connecticut, after leaving the Big Apple.

*Bewitched* was never filmed in Westport, either, but the show did remind TV viewers of the rich history of the New England region by featuring several episodes filmed on location in Salem, Massachusetts, scene of the famous witch trials of 1692, and showing many real historic sites.

There is no real Schooner Bay, Maine *(The Ghost and Mrs. Muir)*, or Danfield, Connecticut *(The Lucy Show)*.

Fairfield, Connecticut, the setting for *Who's the Boss*, is real, but filming for the show did not take place in the town. However, a sweatshirt from the real Fairfield High School did appear on the show.

## BEWITCHED
**(September 1964–July 1972, ABC)**

In this classic sitcom Samantha (Elizabeth Montgomery), a witch, and Darrin (Dick York, and then Dick Sargent), a mortal, settle in Westport, Connecticut, where Samantha has agreed to give up her witchcraft for Darrin and a normal life. But she can't resist twitching her nose a little (her method of conjuring), especially while her husband is at work as a New York ad man. And visits by her witch and warlock relatives don't help the cause.

*Bewitched* was not filmed on the East Coast but mostly on a soundstage at Screen Gems in Hollywood.

In April 1970, however, the set burned down, and in June, the cast hit the road for a spell, filming four on-location episodes in Salem, Massachusetts, and nearby Gloucester, Massachusetts. The Hollywood stage was rebuilt and usable again by July, 1970.

The premise for the series of on-location shows was that Samantha was attending a Witches Convention in historic Salem, scene of the famous witch trials of 1692. Hundreds of citizens of Salem were accused of practicing witchcraft at that time and nineteen people were executed.

Elizabeth Montgomery and her husband (they were later divorced), the show's producer, William Asher, chose the locale, but Dick Sargent seemed to have the best time on location.

During the filming in Salem, Sargent said he particularly enjoyed the attention he received from local fans, who called Sargent and other cast members by their character names.

Sargent really threw himself into the part for an episode filmed in Gloucester, during which Endora (Samantha's mother) turns Darrin into the storm-beleaguered seaman of Gloucester's famous fisherman's statue.

During filming, Sargent was colored a rusty green from head to toe to look like the weather-beaten metal statue. And he remained in costume when the cast took a lunch break at the Gloucester House, a well-known local seafood restaurant.

Proprietor Michael Linquata, noticing Sargent's complexion, asked if anything was wrong with the fish lunch. Montgomery, much amused, answered, "Oh, the meal was wonderful, but it seems as if the fish didn't agree with Mr. Sargent. What do you think we ought to do?"

To which Linquata deadpanned back, "Well, around here, when they come up that color, we generally just hit them over the head and throw them back [into the sea]."

Lucky for Sargent, Linquata's advice was not taken.

The restaurant owner was paid one dollar for a shot of the front of his restaurant used in the episode, and he said local fishermen also made a buck each for appearances as part of the background for the seaside scenes, although most ended up on the cutting room floor, according to Linquata.

In addition to the Gloucester House, the Hawthorne Hotel and several popular area attractions, including the House of Seven Gables, Pioneer Village, and Salem Common in Salem, and Hammond Castle in Gloucester, were featured on the episodes. The hotel hosted a gala witches' brew feast for the cast on the final night of filming in Salem.

## What to See in Salem, Massachusetts

- **House of Seven Gables,** 54 Turner St. The 1668 house immortalized by Nathaniel Hawthorne in his famous book is now open as a museum complex that includes Hawthorne's Birthplace (built in 1750) and other seventeenth-century dwellings. The house was built by Captain John Turner, a merchant mariner. The museum, which also has period gardens and a gift shop, is open daily, 9:30 A.M. to 5:30 P.M., July 1 through Labor Day, and shorter hours off-season. Admission is charged. Combination tickets are available with Pioneer Village: Salem in 1630. Call 508-744-0991.

- **Pioneer Village: Salem in 1630,** Forest River Park. The oldest living history museum in America, Pioneer Village portrays Salem as it existed in 1630, when it was a Puritan fishing village. Costumed interpreters offer guided tours and demonstrate early crafts as well as farm activities. The museum is open Monday to Saturday, 10 A.M. to 5 P.M., and Sunday, noon to 5 P.M., May 29 to October 31. Admission is charged. Combination tickets are available with the House of Seven Gables. Call 508-745-0525.

▶ The **Salem Witch Museum,** Washington Square, offers a multi-sensory presentation about the witch hysteria of the 1690s, featuring thirteen life-sized stage settings. The museum is open daily, 10 A.M. to 5 P.M., with later hours in July and August. Admission is charged. Call 508-744-1692.

▶ **Crow Haven Corner, A Witch Shop,** 125 Essex Street. There are several shops in town specializing in witchcraft supplies, but this is one of the best. The shop is run by Laurie Cabot, who calls herself the Official Witch of Salem. Call 508-745-8763.

▶ **The Hawthorne Hotel,** on the Common. The hotel has eighty-nine rooms and suites with eighteenth-century-style furnishings. A restaurant, Nathaniel's, features American cuisine, and is open seven days a week for breakfast, lunch, and dinner. Live entertainment is offered Thursday, Friday, and Saturday at the hotel's pub, Tavern on the Green, where pub fare is also served. Reservations are suggested. Call 508-744-4080 or 800-SAY-STAY.

▶ **Haunted Happenings** is an annual city festival held around Halloween that attracts a good number of Samantha wannabes. Call the **Salem Chamber of Commerce,** 508-744-0004, for a brochure.

## What to See in Gloucester, Massachusetts

▶ **The fisherman statue,** Stacey Boulevard. Officially called "The Man at the Wheel," the statue was commissioned by the citizens of Gloucester in 1923 to celebrate the seaport's three hundredth anniversary. The inscription on the statue reads, "They That Go Down To The Sea in Ships." The statue was created by local artist Leonard Craske.

▶ **Hammond Castle Museum,** 80 Hesperus Avenue. This medieval-style castle was the home of inventor Dr. John Hays

Hammond, and is open as a museum. The collection includes Roman, medieval and Renaissance artifacts, and the castle is also open to the public for concerts. Open daily for tours, 9 A.M. to 5 P.M. Admission is charged. Call 508-283-2080.

▶ **Gloucester House Restaurant,** Seven Seas Wharf (Route 127). With dining rooms overlooking the harbor, this restaurant fea-

*Bewitched* on location in Gloucester, Massachusetts. In photo are (left to right) Dick Sargent (Darrin), David White (Larry Tate), and Elizabeth Montgomery (Samantha), with restaurant owners Leo and Michael Linquata of the Gloucester House Restaurant, and William Asher, the show's producer. *(Photo courtesy of the Gloucester House Restaurant.)*

tures seafood and American cuisine. Outdoor cafe dining and New England clambakes are offered in the summer. Dinner prices range from $7 to $18. The restaurant is open daily, 11:30 A.M. to 10 P.M. Reservations are advised in season. Call 508-283-1812 or 800-238-1776.

▶ **Seven Seas Whale Watch,** Seven Seas Wharf. Operated by the same people who run the Gloucester House, Seven Seas offers whale-watching expeditions from May 1 to mid-October. Call 508-283-1776.

## Fun Facts

• Among characters appearing on *Bewitched* were Mother Goose, Prince Charming, Julius Caesar, Benjamin Franklin, Queen Victoria, Leonardo da Vinci, Santa Claus (with elves and reindeer), Paul Revere, George Washington (who is charged with disturbing the peace), King Henry VIII, and Napoleon.

• The appearance of famous people on the show is often the result of accidental spells cast by Uncle Arthur or Aunt Clara.

• The December 24, 1970, show was written by the fifth-period English Class at Thomas Jefferson High School in Los Angeles. The show deals with bigotry, and is called "Sister at Heart." All twenty-two kids who worked on the episode received an on-screen writing credit.

• Seven actors played Tabitha on *Bewitched*. On January 13, 1965, she was played by 2 ½-week-old Cynthia Black, and later by 3-month-old twins Heidi and Laura Gentry. Julie and Tamar Young, born June 24, 1965, played 6-month-olds for the first time on February 10, when they were actually 8 months old. Tabitha was later played by Diane and Erin Murphy, from 1966 to 1972.

• In the short-lived spin-off *Tabitha*, the part of a grown-up Tabitha is played by Lisa Hartman *(Knots Landing)*, wife of country singer Clint Black.

• "Durwood" is mother-in-law Endora's pet name for Darrin, who was played by Dick York (1964 to 1969) and later Dick Sargent (1969 to 1972).

• Dick Sargent "came out" and publicly declared his homosexuality in 1991, in protest of a veto of a gay rights bill by California Governor Pete Wilson.

## DARK SHADOWS
**(June 1966–April 1971, ABC)**

While presumably not inhabited by any vampires, the Gothic mansion seen as the home of the Collins family in the opening of each episode of the eerie 1960s soap opera *Dark Shadows* really exists in Newport, Rhode Island. And the scenic city itself is the model for fictional Collinsport, Maine, the small seaside town where 200-year-old vampire Barnabas Collins (Jonathan Frid) lived along with his neighbors, who included ghosts and werewolves.

Despite their idiosyncrasies, the characters on *Dark Shadows* became involved in standard soap opera plots involving lust, romance, and greed. The show attained cult status, with fans including both housewives and teenagers, who would rush home to watch *Dark Shadows* after school in its late afternoon time slot.

The real-life mansion is decidedly less creepy than it appeared on the show. Previously known as Seaview, the 1920s mansion is today called Carey mansion, and is part of Salve Regina University. It's used to house the co-ed Catholic university's music department, with a separate wing used as a dormitory for sixty-three students.

Only exterior shots of the estate were seen on *Dark Shadows*, in which it was called Collinwood. The actors really worked on a reproduction of the mansion's terrace on a set in New York, where the designers added a fountain. Interiors were also shot on the set, and were not replicas of the interior of the real mansion.

Actor Frid visited the mansion as a tourist several years ago and said he got spooked, despite having played a vampire on the show. It seems Frid was caught sneaking a peak through a fence at the property, and a German shepherd approached and started barking.

"I thought I was truly going to be bitten in the environs of the mansion," Frid later said.

No German shepherds prowl the property today, and the Salve Regina University campus is open to visitors.

In the story, Collinwood was supposed to have been built in 1795. The actors portrayed characters both in modern times and in the 1800s.

## What to See in Newport, Rhode Island

▶ **Carey Mansion at Salve Regina University,** Ruggles Avenue (near Bellevue Avenue). Visitors are welcome to tour the campus and view the exterior of the building seen on the show, with its turreted towers, gables, and gargoyles. The mansion was built for Edson Bradley, a liquor baron, and his wife, and was considered an architectural triumph for its time, with much of the stone hauled from the Bradleys' former home in Washington by train and reassembled here. The house gets it name from more recent owners, Martin and Millicent Carey (he is the brother of the former New York governor Hugh Carey), who lease the property to the university.

▶ **The Cliff Walk.** While there is no Widow's Hill behind the mansion in real life, there is the Cliff Walk, a rugged path with the ocean on one side and the rear lawns of the great Newport cot-

The *Dark Shadows* house at Salve Regina University. *(Photo by John W. Corbett, courtesy of Salve Regina University.)*

tages, as the mansions are known, on the other. The unpaved, 3½-mile trail offers embankments of several hundred feet, with views of waves crashing against the rocks.

▶ **The Black Pearl on Bannister's Wharf,** (off America's Cup Avenue). This New England-style eating establishment appeared as the Blue Whale on the show. The restaurant serves New England cuisine with a lunch menu available from 11:30 A.M. to 5 P.M., and a dinner menu from 5 P.M. to 10:30 P.M. Seafood is a specialty. Reservations are required for dinner in the formal dining room in the back of the restaurant, but not for the tavern and bar area. Call 401-846-5264.

To see how the wealthy summer residents of Newport really lived, you can visit one or more of the several mansions open to tourists. For a listing, stop by the Newport Convention & Visitors Bureau's Visitors Center, at 23 America's Cup Avenue. Or call 800-326-6030.

## Fun Facts

- Alexandra Moltke, the actress who played Victoria Winters, figures prominently in Newport history as a major witness in the famous trial of Claus Von Bulow, accused of trying to kill his socialite wife, Sunny, at their Newport mansion. The Von Bulows' cottage, Clarendon Court, is on Bellevue Avenue, down the street from the Carey Mansion.
- Joanna Going, who played Victoria Winters in the 1991 remake of *Dark Shadows,* with Ben Cross playing Barnabas, is from Newport. She told the local newspaper she was not allowed to watch *Dark Shadows* as a child.
- Jonathan Frid (Barnabas Collins) said of the show, "It was a soap opera about the sea as well as monsters." He actually knew the sea quite well, having once served in the Canadian Navy.
- An unaired plot written for the original show had Victoria (Alexandra Moltke) learning that her mother is Elizabeth Collins Stoddard (Joan Bennett). Her paternity was always a secret.
- Kate Jackson *(Charlie's Angels)* played Daphne Harridge. David Selby *(Falcon Crest)* played Quentin Collins. And Louis

Edmonds (Langley Wallingford on *All My Children*) played Roger Collins.

• The part of Barnabas, a vampire, was not originally in the plot of *Dark Shadows*. In fact, the soap was rather traditional at first, involving custody battles and so forth, and was not doing well in the ratings. The writers decided to take some chances, adding characters returned from the dead. Jonathan Frid, who played Barnabas, was originally signed for only five episodes, but became a star of the show.

## I LOVE LUCY
**(October 1951–September 1961, CBS)**

In 1957, the Ricardos moved to Westport, Connecticut, and several references were made on the show to activities and locations around the posh New York City suburb. The show was actually filmed on a set in Hollywood, however.

In the episode called "The Ricardos Dedicate a Statue," the last episode of the Connecticut series, Lucy (Lucille Ball) accidently breaks Westport's famous Minute Men Statue, and tries to replace the artwork by pretending she's the statue herself.

According to Writer Bart Andrews, in *The "I Love Lucy" Book*, for the episode, filmed on April 4, 1957, the show specially commissioned a duplicate of the statue from a studio in L.A.

Other events mentioned as a means of establishing the location in the mind of the viewers included the annual summertime Yankee Doodle Fair, sponsored by the Westport Women's Club.

### WHAT TO SEE IN WESTPORT, CONNECTICUT

▶ **Minute Men statue,** Compo Beach, Compo Road South (off the Boston Post Road). Compo Hill was the scene of a battle between the Continentals and the British in April 1777, and many of those who fell in the battle are buried at the site. Among the Minutemen who survived was Oliver Wolcott, son of a signer of the Declaration of Independence and later Governor of Connecticut. The statue was unveiled on June 17, 1910, by the Connecticut Society of the Sons of the American Revolution.

The Minute Men Statue. *(Photo by Fran Golden.)*

## Fun Facts

• In the Connecticut shows, Mary Jane Croft, who later was Vivian Vance's (who played Ethel) replacement as Lucy's best friend on *The Lucy Show,* played the Ricardo's neighbor Betty Ramsey.

• In a Connecticut episode filmed in 1957, called "Country Club Dance," a twenty-two-year-old Barbara Eden, who later gained stardom on *I Dream of Jeannie,* appeared as a blonde who attracted attention from Ricky, Ralph Ramsey, and Fred.

• "The Ricardos Dedicate a Statue" episode also featured for the first time the real Arnaz children, Lucie, then five, and Desi, Jr., then four. It's the only time they appeared on *I Love Lucy.*

• The final thirteen episodes of *I Love Lucy* were directed by William Asher, who later produced *Bewitched,* starring Elizabeth Montgomery, his wife at the time.

## NEWHART
**(October 1982–August 1990, CBS)**

Opened originally as a tavern in 1810, The Waybury Inn in East Middlebury, Vermont, was chosen by a Hollywood set designer as the quintessential New England inn, thus earning a place in TV history as the setting for the popular sitcom *Newhart.*

On the show, comedian Bob Newhart stars as Dick Loudon, a New York writer of how-to books, who leaves the big city to renovate and run a country inn, the Stratford, with his wife, Joanna (Mary Frann).

In the country, the Loudons come across an odd assortment of characters but have little trouble adjusting to their new life. Dick even takes a job hosting a local Vermont TV show.

The Stratford of the show predates the real inn by several decades, having opened in 1774. And while The Waybury's history is rich and varied, the real inn does not have quite as titillating a past as its fictional counterpart.

In the first episode of *Newhart,* it's revealed that the Stratford may have been more than accommodating to soldiers during the Revolutionary War. According to letters Dick finds, it was known as "the Best Little Inn in New England."

In real life The Waybury Inn was a favorite of poet Robert Frost in the 1950s and early 1960s, and he even had a favorite table in the restaurant.

Just like on the show, The Waybury Inn is run by a married couple. But they are not recently transplanted New Yorkers. Marty and Marcia Schuppert are locals, who bought the inn in 1990.

Other real sites seen on *Newhart* included the nearby town of Middlebury, Vermont, and an historic signpost, actually located in New Hampshire, not Vermont.

For the first few episodes of *Newhart,* the lovely New England shots seen were actually outtakes of the feature film *On Golden Pond,* shot at Squam Lake in New Hampshire, according to Greg Gerdel, director of the Vermont Film Bureau. Scenic Vermont shots were later added.

## What to See in East Middlebury, Vermont

- **Waybury Inn,** Route 125 East. The inn is located in prime leaf-viewing and skiing country. The interior, which is not the same as on the show, includes fifteen guest rooms, all individually decorated, mostly with antiques, and located on the second and third floors. The inn's restaurant has three separate dining rooms and features American cuisine. Breakfast and dinner are served daily, and lunch is offered on a seasonal basis. The inn also serves a popular Sunday brunch. The inn's pub is open until midnight (a pub menu is offered until 10 P.M.), and serves 105 brands of beer. Autographed photos and props from *Newhart* are on display in the inn's lobby and pub. T-shirts that say "Stratford Inn," with a picture of The Waybury Inn, are for sale. Call 802-388-4015 or 800-348-1810.

- ***Across the Fence,*** WCAX, Channel 3. While Dick's *Vermont Today,* was fictional, locals say the closest real-life show in Vermont is this daytime talk show, which features studio guests and independent vignettes about the people and places of Vermont.

The Waybury Inn in East Middlebury, Vermont, as featured on *Newhart. (Photo courtesy of The Waybury Inn.)*

## What to See in Sandwich, New Hampshire

▶ **Colonial sign post,** intersection of Route 113 and Grove Street. The sign indicating mileage to various towns was frequently shown on *Newhart.*

## Fun Facts

• The show was based loosely on BBC's *Fawlty Towers*, staring John Cleese.

• Dick Loudon was the author of do-it-yourself books including *Building Your Own Patio Cover, How to Make Your Dream Bathroom*, and *Care of Your Low-Maintenance Lawn Sprinkler.*

• Tom Poston (George Utley) earlier played Franklin Bickley, a greeting card designer on *Mork & Mindy.* And Newhart later played a greeting card designer turned cartoonist on *Bob.*

• Stephanie (Julia Duffy) and Michael (Peter Scolari) named their baby, born in 1989, Baby Stephanie.

• Darryl (Tony Papenfuss) and Darryl (John Voldstad) spoke only on the final episode of the show, when they both yelled "Quiet!" to their bickering wives.

• In an episode in 1987, Larry (William Sanderson) found out through birth certificates that Darryl No. 1 (Tony Papenfuss) was actually the oldest brother, and Darryl No. 1 took over temporarily as leader of the unusual trio.

• In the show's finale, it was revealed the show was really a dream. Bob Newhart is still on *The Bob Newhart Show,* and wakes up in bed next to his wife on that show, Emily (Suzanne Pleshette). He tells her he had a weird dream about being an innkeeper in Vermont.

• Series star Bob Newhart never stayed at The Waybury, but Larry, Darryl, and Darryl did; their visits included appearances at a nearby camp for terminally ill children, and local fund-raising events.

## PEYTON PLACE
**(September 1964–June 1969, ABC)**

The gossip might not be quite as lush, but there is a real-life town like the one on this show, the first of the prime time soap operas.

Both the television show and the movie, which preceded it, were based on the best-selling book *Peyton Place*, by Grace Metalious. And when the steamy novel was released in 1956, Metalious was living in Gilmanton, New Hampshire, with her husband and three children.

Sleepy Gilmanton quickly became known as the *real* Peyton Place, although the thought makes residents of the town gag, even today.

While exteriors for *Peyton Place*, the television show, were filmed in Hollywood, the movie *Peyton Place* was shot on location in the town of Camden, Maine, about eighty-five miles northeast of Portland.

The producers of the movie moved the setting to the Maine oceanside town because they decided Camden with its picturesque harbor, looked more like New England than inland Gilmanton.

The television show followed the movie's lead, offering a setting designed to reflect New England seaside ambiance.

The town on the show had hitching posts, a bandstand, a pillory, a church, and a wharf, as well as several trees which, according to a *Look* magazine profile of the show, were fakes, with the exception of one live willow tree. The others were California pines that were planted in cement.

According to the article, the town's wharf had real boats, but the show's camera crew made sure to hide from viewers the fact that the boats on the set were actually sitting on dry land.

According to a 1965 *New York Times* story, the studio version of downtown Peyton Place had to be shot close up so as not to show the real high-rise apartment development that loomed above the set.

The real town of Gilmanton, population 3,000, is located in the Lakes Region of New Hampshire near Laconia. It's a bedroom community, with people commuting mostly to Manchester or Concord.

Author Metalious, whose *Peyton Place* novel sold 20 million copies, frequently angered her Gilmanton neighbors with her comments about the town.

"To a tourist," she once said, "these towns look as peaceful as a postcard. But if you go beneath that picture, it's like turning over a rock with your foot—all kinds of strange things crawl out."

Metalious herself was vulnerable to drinking binges and infidelity. Today the only visible reminder of Metalious in Gilmanton is her gravestone. Said Betty Smithers, long-time town clerk of the town, when asked of Metalious's legacy, "We're not proud of it."

The Camden, Maine, waterfront; the model for the seaside setting of *Peyton Place*. *(Photo courtesy of the Maine Office of Tourism.)*

## What to See in Gilmanton, New Hampshire

▶ **Grace Metalious grave,** Smith Meeting House Cemetery, on Smith Meeting House Road (off Route 140). The author died of liver-related illness at age thirty-nine, on February 25, 1964, seven months before the show premiered.

## What to See in Camden, Maine

▶ **The Owl and Turtle Bookshop,** 8 Bay View. When asked if Constance MacKenzie's shop on the show was modeled after a real downtown Camden bookstore, the Chamber of Commerce pointed us here. But this bookshop actually didn't open until 1970, a year after *Peyton Place* ended its five-season run. Call 207-236-4769.

## Fun Facts

• Mia Farrow (Allison MacKenzie), whose real life became a soap opera co-starring Woody Allen, Frank Sinatra, and Andre Previn, was making headlines even in her *Peyton Place* days. In the summer of '65, for instance, she took a several-week-long cruise off the New England coast with Frank Sinatra on his boat *Southern Breeze*. During that time, her character on the show conveniently had a car accident and was in a coma, and a stand-in actress was used for the hospital scenes.

• In the *Peyton Place* heyday, during the 1965–1966 season, episodes ran three days a week. They were reduced to two by the 1966–1967 season.

• The show was purposely less sordid than the book, and the Cross family of the book was not even on the show.

*Look* magazine suggested the following rules for watching:

1. If you suspect anything, you're wrong.
2. You need to watch every episode.
3. If a villain appears, he's probably not one.

• Actress Dorothy Malone (Constance) was ill for a time with life-threatening blood clots necessitating chest surgery, and was substituted by actress Lola Albright.

• Jack Paar, of *The Jack Paar Show,* visited the real Gilmanton to interview residents about *Peyton Place* in 1965. Paar found mixed reviews for the show in town. But he also found that everyone watched the show.

• Ted Post, the director of the show, once jokingly offered fans a do-it-yourself kit to find out what happens next. It had cut-outs of fifteen main characters, a roulette wheel with plot twists, a pack of emotion cards, a "get out of jail free" card, and a copy of Erich Fromm's *The Art of Loving*, reported *Insider's Newsletter for Women.*

On June 4, 1965, ABC offered the following explanation of the main characters:

• Constance Carson (Dorothy Malone), owner of the town's book shop, who after eighteen years of guiltily hiding her secret, marries Elliott Carson, father of her daughter, Allison.

• Allison MacKenzie (Mia Farrow), the seventeen-year-old daughter of Constance and Elliott, who retains her legal surname. She's a freshman at Peyton College.

• Elliott Carson (Tim O'Connor), father of Allison MacKenzie. He has recently been found innocent of the murder of his first wife and pardoned after eighteen years in prison. Marries the former Constance MacKenzie.

• Betty (Barbara Parkins) a nurse's aid seeking to establish a life of her own and forget her marriage to Rodney and the miscarriage she suffered.

• Rodney Harrington (Ryan O'Neal), older son of Leslie and the late Catherine Peyton Harrington, and grandson of Martin Peyton of Boston, owner of Peyton Mill, the town's biggest enterprise. Briefly married to Betty. His father had the marriage annulled.

• Dr. Michael Rossi (Ed Nelson), a New York bachelor internist who recently settled in town and wants to get to know his patients.

• Leslie Harrington (Paul Langton), father of Rodney and a town native from the wrong side of the tracks who married the daughter of the richest man in town. After his wife's death, he is removed as head of the mill.

## WINGS
**(April 1990– , NBC)**

*Wings* was created by three *Cheers* producers, David Lee, Peter Casey, and David Angell, who originally said they did not want the show to look like a protégé of its predecessor. And they especially didn't want the setting located down the street from Boston's *Cheers*.

The producers, in fact, searched through the Pacific Northwest for a small airport on which to base the show. But finding none with the right ambiance, they headed to New England.

Angell, who grew up around Providence, Rhode Island, suggested they look at Provincetown, Martha's Vineyard, and Nantucket. The producers settled on Nantucket because it offered the right amount of sedate charm. Yes, despite what people on the West Coast might think, people in the East can be laid back too.

Reviewers were quick to point out the similarities between *Wings* and *Cheers*, which, in addition to geography, include the fact that the sets for the two comedy shows look a bit alike. On *Wings* there is not a bar, but a lunch counter and tables, around which a lot of the action is centered. And both feature a colorful collection of regular characters.

Nantucketers are generally proud that some exteriors for *Wings* are really filmed here on their sleepy island, and merchants are happy with the extra attention the show has brought.

The actors actually do their thing on a set in Los Angeles. But that's not to say that the show hasn't occasionally caused a ruckus on the island.

In one episode, for instance, spacey Lowell (Thomas Haden Church) climbs to the top of a church steeple for some deep thinking. Believing Lowell is planning to jump off, Joe (Timothy Daly) climbs up to rescue him. The scene was filmed at Nantucket's Unitarian Universalist Church, using stand-ins for the actors. But not everyone on the island got the word. And a crowd gathered thinking a real jumper was on the church's steeple.

The show's introduction features exterior shots of the real Nantucket Memorial Airport, but interiors are all done in Los Angeles, and the inside of the real airport does not look like the one on the show. Also shown on various episodes have been a local video store, downtown Nantucket, several island houses, local restaurants, and the Sankaty Head Light House, according to island officials.

Nantucket at Christmastime, when the cobblestoned Main Street is impressively decked out with dozens of Christmas trees as part of the annual Nantucket Noel and Christmas Stroll events, has also been featured.

Local merchants, making the most of the show's success, sell *Wings*-themed items featuring a large winged design at various venues around town. Not coincidentally, the items are marketed by Bull & Finch Enterprises, the same firm that markets *Cheers* memorabilia.

## What to See in Nantucket, Massachusetts

▶ **Nantucket Memorial Airport,** 30 Macy's Lane. Located southeast of Nantucket town, the island's real airport is one of the busiest airports in Massachusetts, offering service to Boston, New York, and other cities, on both a regularly scheduled and charter basis. The airport building was recently renovated, but

The tower at Nantucket Memorial Airport, seen on *Wings*. *(Photo by Rob Benchley.)*

retains a tower that looks in real life as it is shown on the show. The real building's new decor is described by officials as being very pristine, featuring columns and gleaming counters that might remind visitors of a bank building, but not of the set of *Wings*. The airport has a restaurant, Hutch's, that is located in a separate room, not in the lobby as on the show.

For those wanting a flying *Wings* experience, flight instruction is available. Call 508-325-5300.

▶ **Nantucket Windmill Gifts,** Nantucket Memorial Airport, offers for sale *Wings* items including T-shirts, sweatshirts, and hats. Other shops carrying *Wings* items can be found at **Straight Wharf,** where the ferries from Hyannis dock.

▶ **Unitarian Universalist Church,** 11 Orange Street. Scene of a scene, as described above.

## Fun Facts

• On *Wings*, the airport is named Tom Nevers Field. Tom Nevers is actually the name of a residential area on the island. No one in town seems to know who Tom Nevers was for sure, but he's believed to have been a native American who lived on the island in colonial times.

• Timothy Daly's big sister is Tyne Daly, of *Cagney & Lacey*.

• *Wings* was the first TV show to sign the Coke body guy, Lucky Vanous (he appeared with his shirt off in a Coke commercial playing a construction worker being admired by female office workers), for a guest appearance.

• Crystal Bernard, who plays cello-playing coffee shop proprietress Helen Chappel, rides a Harley in real life and has raced in pro-celebrity car races. She is also a songwriter. She earlier appeared on *It's a Living*.

# MANHATTAN

It's big! It's loud! And it offers a little bit of everything. So it's no wonder New York City has attracted a myriad of TV productions.

Sitcoms and cop shows seem to be the Big Apple's biggest draw, with a wacky redhead, a neurotic comedian, an obstetrician, another redhead (this one a cop), and a lollipop-sucking police lieutenant among New York's most popular TV creations.

A lot of the locations mentioned on the New York shows are not real, the Twelfth Precinct on *Barney Miller*, among them.

And the address given as the location of Lucy and Ricky's home on *I Love Lucy* (623 East 68th Street) would actually put the duo, and visitors who try to find the apartment building, in the East River.

Reality did not escape the Hollywood producers of the 1960s show *Naked City*, however, who looked at both the grittiness and elegance of the real city by shooting on location in the Big Apple. *Law & Order*, *The Equalizer*, and Fox's *TriBeCa* followed suit, also showing the city as more than passing scenery.

But reality is not always the way it seems.

*Kojak*, a show that brought New York cops into an international spotlight, was really based for three years in L.A., doing much of its filming there, before it headed to New York for extensive location work.

And New York cop shows *Cagney & Lacey* and *NYPD Blue* only really visited New York a few times a year, filming so as to trick viewers into thinking Big Apple. Both shows really shot most of their action in Hollywood.

Even *Seinfeld*, a show that is all about the mundane aspects of life in Manhattan, calls a Hollywood set its home. And the show unabashedly uses shots of an L.A. apartment building as Jerry's New York home (although a real New York coffee shop is shown as his hangout).

But such charades are nothing new.

Ann Marie (Marlo Thomas), on the 1960s sitcom *That Girl,* visits New York's real Sardi's Restaurant, and looks in the windows of the real Bergdorf Goodman on Fifth Avenue; and Rhoda Morgenstern (Valerie Harper), on *Rhoda,* catches a subway at the real West 72nd Street station on her wedding day. But both shows were Hollywood-based productions.

Also of the Hollywood-based lot is the big hit of the 1994–1995 season, *Friends.* The show does feature the outside of a real New York apartment building, however. It's at 90 Bedford Street in the West Village.

*The Cosby Show, Kate & Allie,* and *The Days and Nights of Molly Dodd* were done in New York studios, but did not feature all that much of the city's real streets, either.

TV shows that did show the sights and sounds of New York included visiting shows like *The Beverly Hillbillies,* which had the Clampetts coming to the Big Apple for three episodes. The obvious out-of-towners took in the United Nations, the Staten Island Ferry, and Central Park (where they met Sammy Davis, Jr., in a walk-on role as a Central Park cop).

## CAGNEY & LACEY
**(March 1982–August 1988, CBS)**

This ground-breaking New York cop show featured two female cops, one married with kids and the other single. All three networks originally turned the series down. But the show's pilot, shown as a TV movie, got a lot of attention and CBS took a chance.

Cagney (Sharon Gless) and Lacey (Tyne Daly) lived through many travails on the show: Chris Cagney faced date rape, the death of her father, and alcoholism, and Mary Beth Lacey dealt with family life and working mom issues, both while handling their load as detectives at the Fourteenth Precinct in New York.

Despite its New York look and feel, the show was filmed mostly in L.A.

At the beginning of each season, the stars were brought to New York for a week, filming at places to establish the Big Apple as the show's setting, such as Rockefeller Center, Brooklyn, Times Square, SoHo, and Central Park.

"The exterior scenes are always inserted early in the story," Barney Rosenzweig, the show's producer explained, "in order to establish New York in the viewer's eye and mind. Then, as the story goes along, we trust that the strength of the story will so involve viewers that they stop thinking about where it was filmed."

Among the tricks used were affixing New York newspaper logos on delivery trucks in L.A.

"Tyne and Sharon might act a scene in front of that truck, but some savvy viewers know that if the New York truck drove

The New York Supreme Court Building. *(Photo by Fran Golden.)*

away, you'd see it had been covering up a palm tree," Rosenzweig said.

The show was canceled in the spring of 1983, but topped ratings that summer in reruns. Viewers supported the show with letters, and it was brought back in the spring of 1984, after a six-month hiatus.

## What to See in Downtown Manhattan

▶ **The New York Supreme Court Building,** 60 Center Street. One of the most frequently photographed buildings in New York, this block-size courthouse first appeared in the movie *Miracle on 34th Street,* and has since been seen in many movies and TV shows, the latter including *Cagney & Lacey, Kojak, NYPD Blue,* and *Law & Order.* Most of the TV productions use the building's exterior and shoot inside in Room 452, which is used as a jury assembly room in real life, but is identical to other courtrooms in the building. The courthouse is one of the busiest in the world. The entrance and rotunda areas have 1930s murals depicting city scenes and New York history; the artwork had been in a state of disrepair but was recently being restored.

## Fun Facts

• In 1982, there were 1,400 women on the New York City police force of 18,000, and 95 were detectives. Women first went on patrol in the city in 1973.

• The role of Cagney was originally played by Loretta Swit (in the pilot movie), and then by Meg Foster. Sharon Gless took over the part in 1983, and at the time there were rumors that the reason for the replacement was that Tyne Daly and Foster depicted what was considered to be "too masculine" a pairing. According to a March 1982 article in the *Daily News,* at one point it was suggested that Raquel Welch and Ann-Margret play Cagney and Lacey.

• Cop Mary Beth (Tyne Daly) was arrested herself in one episode of the show, for protesting the transport of nuclear waste.

• Tyne Daly won Emmys for outstanding actress in a drama series for her portrayal of Mary Beth Lacey in 1983, 1984,1985, and 1988. And Sharon Gless won for her portrayal of Cagney in 1986 and 1987.

• Producer Barney Rosenzweig, who formerly worked on *Charlie's Angels*, reportedly developed the idea for the show after reading Molly Haskell's book *From Reverence to Rape*, about the treatment of women in film.

• Josie the bag lady was played by Jo Corday, Barney Rosenzweig's mother-in-law in real life at the time.

• Barney Rosenzweig and Sharon Gless were married in May 1991. Rosenzweig at the time was producing *The Trials of Rosie O'Neill* (September 1990–December 1991, CBS), an attorney show starring Gless.

• After *Cagney & Lacey*, Tyne Daly starred in *Gypsy* on Broadway, winning a Tony award for her performance in 1990.

• Sharon Gless and Tyne Daly were reunited as the cop duo in two TV reunion movies. While the first movie, which aired in 1994, was called *Cagney & Lacey: The Return*, the actresses, according to *TV Guide*, referred to the endeavor as *Cagney and Lacey: The Menopause Years*.

## THE COSBY SHOW
**(September 1984–April 1992, NBC)**

This hit sitcom, the top show of the 1980s, was comedian Bill Cosby's baby. Dr. Cosby (he has a Ph.D. in education) co-created, produced, and wrote episodes of the show, which featured Cosby's own brand of family humor (he even cowrote the theme song).

The perspective of the show was the dad's—that of Cosby as Heathcliff "Cliff" Huxtable, an obstetrician married to a lawyer, Clair (Phylicia Rashad). Together, and with a good dose of humor (and many smirking close-ups of Cosby), Cliff and Clair raised their five kids, four girls and a boy, in an upscale Brooklyn brownstone (Cliff's office is downstairs). And they later shared the joy of getting to know their very cute grandchildren as well. They were a big happy family.

*The Cosby Show* received both criticism and accolades for the fact that the parents were happily married black professionals (the critics said the family was not very real).

But the message of the show was family love, not black or white.

While sibling rivalry and teenage woes were frequently topics of discussion, as was Cliff's battle with weight gain, education was stressed on the show, as well as traditional values. Viewers could watch the show with their own families and see something of themselves. The plots also dealt sensitively with current topics.

For example, exiled South African singer Miriam Makeba guest-starred as young Olivia's (Raven-Symone) grandmother in 1991, gently explaining to the child, and to viewers, apartheid in South Africa.

An episode created with the input of Landmark College, a school specializing in teaching students to deal with the learning disability dyslexia, educated viewers by showing Theo's (Malcolm-Jamal Warner) efforts at dealing with his own dyslexia.

In the finale of *The Cosby Show,* the several generations of Huxtables, grandparents, parents, kids, and their kids, were on hand to watch Cliff and Clair dance offstage. Denise (Lisa Bonet), who was living in Singapore, called to announce she was pregnant.

And in the final episode, the pesky doorbell was finally fixed.

Since Cosby, who was raised in Philadelphia, reportedly did not particularly like Hollywood, the show was taped instead at studios in New York.

## What to See in Greenwich Village

▶ **The private home,** 10 St. Luke's Place (off Seventh Avenue South). The facade of this Greenwich Village brownstone was used as the Huxtable's house and the block itself was also featured. On the show, the family lives at 10 Stigwood Avenue, in Brooklyn Heights.

## Fun Facts

• The character of Sondra (Sabrina Le Beauf), the oldest Huxtable daughter, was added after the show's pilot (on which Cliff and Clair had four kids). While she was referred to on

The private home seen on *The Cosby Show.* *(Photo by E. L. Tobias.)*

earlier episodes as attending Princeton, she did not appear in person until November 1984.

• Actress Lisa Bonet, who played rebel Denise, was pregnant in 1988, by her then-husband, rock star Lenny Kravitz. Her character was not, so the pregnancy was hidden.

• The show was the first to rank number one for four years in a row, from 1985 to 1989.

• Phylicia Rashad was Phylicia Ayers-Allen when the show started, but NBC sportscaster Ahmad Rashad romantically proposed to her on one of his broadcasts, and they were married. She is the sister of choreographer/dancer/actress Debbie Allen *(Fame)*.

• The theme song, cowritten by Bill Cosby and Stu Gardner, was originally played by jazz greats including Grover Washington, Jr., then by a cappella singer Bobby McFerrin, then by the Oregon Symphony. The choreography in the intro was by Geoffrey Holder.

• On the show's spin-off, *A Different World,* Denise (Lisa Bonet) headed off to the fictitious Hillman College in Georgia, which was also the alma mater of her father and grandfather. Bonet left the show after a year, returning to *The Cosby Show* for occasional appearances.

• Bill Cosby earlier costarred with Robert Culp in *I Spy.* He had two other series that were not as successful, *The Bill Cosby Show* and *Cos,* but did well with his cartoon creation *Fat Albert and the Cosby Kids.*

• The original proposal for *The Cosby Show* called for the father to drive a limo and the mother to be a plumber. It was rejected by the networks. Legend has it that ABC, and possibly CBS too, later rejected the final proposal for the hit show too.

• Bill Cosby has won five Grammy awards, at last count, for his comedy albums.

• Malcolm-Jamal Warner, who played Theo, is a middle-class kid who tried to do his audition in jive. Bill Cosby told him to be himself, and he got the part.

• Stevie Wonder guest-starred on the show, singing "I Just Called to Say I Love You" with Clair.

• The show created a fashion trend. Bill Cosby's loud sweaters became a favorite holiday gift for dads.

## DREAM ON
**(July 1990– , HBO)**

Martin Tupper (Brian Benben) is a TV child of the 1950s on this unusual and hilarious R-rated sitcom. Martin deals with the hassles of life and relationships in Manhattan by remembering scenes from old shows and movies (real clips are shown to reveal his inner thoughts).

The comedy is clearly adult, and the racey show benefits by appearing on a cable channel (without network censorship). Martin, who is single, likes sex, and he has it frequently with a variety of women, many of whom bare their breasts. He also uses four-letter words.

The apartment building where Martin lives on *Dream On.* *(Photo by Fran Golden.)*

The Court Square Building. *(Photo by Fran Golden.)*

The character still has a thing, of sorts, for his ex-wife, Judith (Wendie Malick), who was married to the greatest man on earth, Dr. Richard Stone (who is never actually seen on the show).

Stone and Judith have a superbaby son, who is called Little Richard. Martin and Judith have a less prodigious but definitely horny teenage son, Jeremy (Chris Demetral).

The show features New York scenes, but is actually filmed in a studio in L.A.

## What to See in SoHo

▶ **The Apartment Building,** 196 Spring Street (at Sullivan). This SoHo apartment building is the real-life location seen as Martin Tupper's home on the show.

▶ **The Court Square Office Building,** 2 Lafayette Street (across from the U.S. Court House). This is where Martin Tupper works at the publishing house, but there aren't any publishers here in real life. The building actually houses city offices.

## Fun Facts

• Denny Dillon, who plays Martin's frustratingly ditzy (or is she just mean?) secretary, Toby Pedalbee, previously appeared on *Saturday Night Live.*

• Michael McKean, who plays boss Gibby Fiske, starred in the movie *This is Spinal Tap,* and was Lenny on *Laverne & Shirley.*

• In 1994, at age forty-six, Michael McKean was also hired to replace Phil Hartman on *Saturday Night Live,* which he did in addition to *Dream On.* He told *Entertainment Weekly* that he wanted to try something new. "And hang gliding was out, so SNL was in," McKean told the magazine.

• Brian Benben, who plays Martin, is married in real life to movie actress Madeleine Stowe *(Bad Girls, Last of the Mohegans.)*

• The police squad in a 1994 kidnapping episode was made up of actors who earlier played cops on *Hill Street Blues,* including Ed Marinaro, Joe Spano, Michael Warren, Taurean Blacque, and Bruce Weitz.

• Movie director John Landis is the show's executive producer.

• Martin's longtime best buddy, Eddie Charles (Dorien Wilson), told their high school classmates Martin was gay, so Eddie could get the girls himself.

• Saintly Richard Stone, Judith's husband, is killed off in 1994 and is mourned at a service at the United Nations as part of an "International Day of Sorrow" in his honor that happens to fall on Martin's own birthday.

• Actress Wendie Malick frequently guest-stars on network shows, including an appearance as a battered wife on an episode of *NYPD Blue*.

## FAME
**(January 1982–August 1983, NBC, then syndicated)**

The real High School for the Performing Arts in New York was the setting for the plot of this drama/music/dance show, which dramatized the lives of students at the school.

The old High School for the Performing Arts building. *(Photo by Fran Golden.)*

Singers, dancers, and musicians were featured and the show gained cultlike status among some viewers, particularly those in Europe.

*Fame* was mostly shot on a Hollywood set, but exteriors of the real New York high school, which has since moved to a new location, were also featured.

Erica Gimpel, who played fame-seeking singer-dancer Coco Hernandez, was a real New York High School for the Performing Arts student who took a leave from school to appear in the show.

### What to See in Manhattan

▶ **The old High School for the Performing Arts building,** 120 West 46th Street. This is the school that was shown on the show, but the city's high school performing arts program has since moved, and the building is now used for other alternative school programs.

▶ **High School for the Performing Arts,** LaGuardia High School, 108 Amsterdam Avenue. This is the new location for high school students studying the performing arts in New York.

### Fun Facts

- The show was based on the movie *Fame,* which won Oscars for Best Song and Best Original Score in 1980.
- Three stars from the movie *Fame* were also on the show, Gene Anthony Ray as Leroy, Albert Hague as music teacher Mr. Shorofsky, and Lee Curreri as musician Bruno Martelli.
- Debbie Allen played talented dance teacher Lydia Grant, and Allen really did choreograph the show's dance numbers. She also continued to gain fame as a choreographer after the show's run. Allen is the sister of Phylicia Rashad of *The Cosby Show.*
- Lee Curreri actually wrote some of the show's music.
- The theme song "Fame" was written by Michael Gore and Dean Pitchford, and sung by Erica Gimpel, who played Coco.
- Janet Jackson appeared on the show for a year as Cleo Hewitt.

• Three albums based on the show became hits in England. The cast of the show toured as The Kids from *Fame*, in England, as well as in Israel, Holland, Belgium, Sweden, and Finland.

• Carrie Hamilton, who played needy Reggie Higgins, is Carol Burnett's daughter.

• Art Carney *(The Honeymooners)* guest-starred on the show with Ray Walston *(My Favorite Martian)*, as a janitor and his aging vaudevillian pal.

## THE JEFFERSONS
**(January 1975–July 1985, CBS)**

George and Louise Jefferson, who used to be Archie Bunker's neighbors *(All in the Family)*, move on up to the East Side to a luxury high-rise apartment with four bathrooms in this spin-off show that also became a popular sitcom.

Sherman Hemsley starred as tightly wound George, who runs a successful string of dry cleaning establishments (including one in the lobby of the apartment building where he lives). And Isabel Sanford played his patient wife Louise, or Weezie, as George calls her.

George is just as much a bigot as Archie, and for almost three years on *All in the Family*, George was referred to but not seen. Isabel explained her husband was prejudiced against whites and refused to visit the Bunkers.

In a 1978 episode of *The Jeffersons*, George and Louise are tied up by a burglar and reminisce in flashbacks of their days in Queens as the Bunkers' neighbors.

*The Jeffersons* introduced an interracial TV couple, rare at the time, Tom (Franklin Cover) and Helen Willis (Roxie Roker), who live upstairs from the Jeffersons. The Willises' daughter, Jenny (Berlinda Tolbert), eventually marries the Jeffersons' son, Lionel (Mike Evans, and then Damon Evans).

*The Jeffersons* was taped in Hollywood before a studio audience, but the outside of a real East Side apartment building is shown as the family's home in the show's intro.

## What to See in Manhattan

▶ **The high-rise apartment building at 85th Street and Third Avenue.** This large East Side apartment building is shown as the place the Jeffersons moved to when they did their "movin' on up, to the East Side, to a deluxe apartment in the sky . . ."

## Fun Facts

• The intro song "Movin' on Up" was by Jeff Barry and Ja'net DuBois. She later played Willona on *Good Times*.

• Isabel Sanford is from New York and appeared in the movie *Guess Who's Coming to Dinner.*

*The Jeffersons* "moved on up" to this high-rise apartment building. *(Photo by E. L. Tobias.)*

• The part of Lionel was originally played by Mike Evans. The actor left to work on *Good Times*, which he co-created, and was replaced by Damon Evans (no relation), who had appeared off-Broadway in *Godspell*. Mike Evans returned to the show, and the role of Lionel, in 1979.

• Marla Gibbs, who played Florence Johnston, the Jeffersons' outspoken maid, starred in the show's very short-lived spin-off *Checking In*.

• Like *All in the Family, The Jeffersons* was produced by Norman Lear.

• Actor Sherman Hemsley went on to play another outspoken TV lead role, as the deacon star of the sitcom *Amen*.

## KOJAK
**(October 1973–April 1978, CBS)**

Inspired by *The Marcus-Nelson Murders*, a CBS TV movie in 1973, and based on a real 1963 murder case involving two young career women in New York, *Kojak* was an immediate hit, making the top ten list of shows its first season.

Telly Savalas, who earlier acted in some sixty movies, played Lieutenant Theo Kojak of the Thirteenth Precinct, Manhattan South, a gruff, bald, quick-witted, lollipop-sucking investigator, who had already worked on the force for 20 years when the show began. "Who loves ya, baby?" became the character's, and the actor's, catch phrase.

The show was applauded by police for its realistic depictions of crime in the big city, and Savalas became a darling of the police set, receiving a plaque from the New York Police Commissioner and honors from the Society of Professional Investigators. The PTA (Parent Teacher Association), however, was not so crazy about *Kojak*, putting the show on its list of the worst ten TV shows for its violent content.

The cop show was initially filmed mostly in Hollywood, but started filming extensively in New York starting in 1976, with the producers saying location filming would add dramatic impact.

CBS touted the move as allowing *Kojak* to show New York "from high-rise East Side apartments to derelict warehouses,

New York's Ninth Precinct, as featured on *NYPD Blue* and *Kojak*. *(Photo by E. L. Tobias.)*

from the Port of New York to Kennedy International Airport, the United Nations and Greenwich Village."

George Savalas, brother of Telly, and Detective Stavros on the show, told the New York *Daily News*, "You can't buy for all the money in the world the stuff you have in New York. How are you going to fake the Fulton Fish Market or steaming sewer caps and the hustle and bustle of New York?"

## What to See in Manhattan

▶ **The Ninth Precinct Building,** 321 East 5th Street (between First and Second Avenues.) Home-base to some three hundred cops, the East Village precinct building was featured as the Thirteenth on *Kojak*. Exterior shots of the building were used, and some real rooms inside were duplicated on the show's set. The precinct building is also featured on *NYPD Blue*, where it is known as the Fifteenth.

## Fun Facts

• Actor Telly Savalas, whose real first name is Aristotle, first shaved his head for the role of Pontius Pilate in the movie *The Greatest Story Ever Told*.

• "Crock a-a-ah," was Kojak's frequent call to his flunky, Detective Bobby Crocker (Kevin Dobson).

• A Florida teenager, Ronald Zamora, tried to use the show in his defense after he murdered his eighty-two-year-old next-door neighbor. His lawyers wanted to claim the show had caused the youth to go crazy, but the argument was rejected, as was a request to call Telly Savalas to the stand. Zamora got life in prison.

• The Savalas brothers as well as actors Dan Frazer, who played Frank McNeil, and Kevin Dobson, who played Bobby Crocker, were all from New York City.

• George Savalas used his middle name, Demosthenes, in the credits during the show's first few years.

• Telly Savalas majored in psychology at Columbia University and once worked in the information section of the U.S. State Department. He released at least one record album that did well in Europe, and sang at an Emmy broadcast.

• Telly Savalas returned to the role of Theo Kojak for ABC in 1989, starring in several Saturday Night Mystery movies. Kojak's position was elevated to inspector in the movies, and Bobby Crocker, in a 1990 episode, was an assistant DA. Actor Dobson, at the time, was a regular on *Knots Landing*.

• Singer Paul Anka made his acting debut on *Kojak* in 1974, playing a stoolie who was using a detective to further his sleazy career.

## LAW & ORDER
**(September 1990– , NBC)**

This realistic cop/law drama follows the legal process in two stages, from the police side, with the investigation and the arrest, and as cases proceed through the criminal courts.

The show's stars keep changing, but a constant on the show is its real New York aura. The show is shot entirely on location in New York, and co-executive producer Joseph Stern told *The New York Times* that the city is always there in the background, if not the foreground. Rather than focus on New York landmarks to establish locale, the producers seek to capture the real sights and sounds of the city, Stern said, like a glimpse of a street scene outside a window, or the sound of traffic.

Crews from the show shoot all over the city, using as their base a converted pier (at West 23rd Street and the Hudson River), and shooting some interior scenes at space in a former city courthouse (lent to the production by the mayor's office, which now occupies the building).

The Tweed Courthouse. *(Photo by Fran Golden.)*

To understand police work, actor Chris Noth, who plays young Detective Mike Logan, tagged along with New York's real Thirty-fourth Detective Squad.

The show finds many of its storylines in tabloid headlines, then seeks to deal with the issues on an intelligent level.

The real 1990 fatal fire at the Happy Land Latino social club, racial troubles in Bensonhurst, the Mayflower Madam scandal, a rape case at St. John's University, and an abortion clinic bombing all have inspired episodes.

An impressive ensemble cast plays the cops and lawyers.

## What to See in Downtown Manhattan

▶ **The Tweed Courthouse,** 52 Chambers Street. Once a courthouse, this large building now houses city offices, including those attached to the mayor's office. The outside is seen on *Law & Order,* and production for the show also takes place in some appointed offices inside (which are not open to the public).

## Fun Facts

- The concept for *Law & Order* is actually an update of a 1960s ABC show *Arrest and Trial,* which starred Ben Gazzara and Chuck Connors as a cop and an attorney doing their jobs in L.A. In the earlier show, the attorney was on the defense side.
- Actor Paul Sorvino, who plays robust Italian Detective Sergeant Phil Cerreta, was also an opera singer, and he left the show to pursue that talent. He was also a real deputy sheriff in Scranton, Pennsylvania and an auxiliary policeman in Rye, New York.
- When George Dzundza, who plays Detective Sergeant Max Greevey, left the show, his character was killed off.
- When actor Michael Moriarty left the show, his character, Assistant DA Benjamin Stone, quit his job, frustrated after a witness is killed. Moriarty earlier won Emmy's for his performances opposite Katharine Hepburn in *The Glass Menagerie,* and in the miniseries *Holocaust.*
- Dick Wolf, creator and co–executive producer of the show, was earlier a producer for *Miami Vice* and script consultant on *Hill Street Blues.*

• Law & Order was the highest rated new drama in 1990.

• Advertisers balked at episodes dealing with abortion, the Irish Republican Army, and child abuse, pulling ads.

• Joining the open-door ensemble cast in 1994 was Jerry Orbach *(The Law and Harry McGraw)*, as a cop, and Sam Waterston *(I'll Fly Away)*, as a DA.

## MAD ABOUT YOU
**(September 1992– , NBC)**

Comedian Paul Reiser, who stars as Paul Buchman in this hit sitcom, also cocreated, writes, and produces for the show, which deals with a couple's attempt to find marital bliss in the 1990s.

Reiser said in numerous interviews that a lot of the material for the show comes from his own real-life marriage to Paula Reiser, a psychotherapist. Other cast and crew members and cocreator/executive producer Danny Jacobson also chip in details from their own marriages, Reiser has said.

Reiser's character, Paul, is a filmmaker, and Helen Hunt plays Paul's wife, Jamie who used to work in public relations.

The show's themes are geared toward an adult audience. A long-delayed night of romance planned by Paul and Jamie in one 1993 episode, for instance, is frustratingly interrupted by phone calls, arguing neighbors, joggers, and the couple's dog, Murray.

In another episode, model Christie Brinkley and tennis champ Andre Agassi guest-starred as the couple's fantasies. Paul and Jamie think of the stars while hooked up to a virtual reality simulator.

Reiser himself grew up in Manhattan, in Stuyvesant Town, and even though the show is done in Hollywood, some real locations in Manhattan are shown.

### What to See in Manhattan

▶ **The apartment building** at 12th Street and Fifth Avenue (on the southeast corner). This is the building shown as the location of Paul and Jamie's home on the show.

▶ **The Old Town Bar & Restaurant,** 45 East 18th Street. This landmark restaurant appears as a bar on the show. David Let-

Paul and Jamie reside in this apartment building on *Mad About You.* *(Photo by Fran Golden.)*

The Old Town Bar & Restaurant. *(Photo by E. L. Tobias.)*

terman also used the restaurant as a location. Burgers, salads, and sandwiches are specialties. The restaurant is open Monday to Friday, 11:30 A.M. to 3 P.M., for lunch, and 5 P.M. to 1:30 A.M., for dinner. On Saturday it's open 1 P.M. to 11:30 P.M., and on Sunday, 3 P.M. to 10 P.M.

### FUN FACTS

- Helen Hunt (Jamie) earlier appeared on *The Mary Tyler Moore Show* as Murray's (Gavin MacLeod) daughter.
- Paul Reiser (Paul) earlier co-starred in *My Two Dads*. He also has appeared in movies, including *Diner* and *Aliens*.
- Comedian Reiser is a friend of *Seinfeld*'s Jerry Seinfeld (they were stand-up comics in New York together), and as a tradition they get together with other comic friends every New Year's Day.
- In 1994, Paul Reiser published a book called *Couplehood*, which focuses, like *Mad About You*, "on a couple and the little moments, the tiny beats between them," Reiser told *TV Guide*.
- In a flashback sequence in 1994, it is revealed that Paul proposed to Jamie at the ice-skating rink at Rockefeller Center on Valentine's Day.

## NYPD BLUE
**(September 1993–      , ABC)**

Cocreators Steven Bochco *(Hill Street Blues, L.A. Law)* and David Milch *(Hill Street Blues, Capital News)* got our attention with a nude butt scene (actress Amy Brenneman's, in the premiere episode).

But despite gasps from conservatives, this New York cop show for adults quickly drew loyal fans and, in 1994, received a record twenty-six Emmy nominations.

The show includes some violence, a touch of nudity, and use of expressions like "scum bag," and other language not usually heard on prime time TV. Just in case viewers are unaware of its adult content, the show opens with the disclaimer that "this police drama contains adult language and scenes with partial nudity. Viewer discretion is advised."

Reverend Donald Wildmon of the American Family Association ran an ad in major newspapers suggesting sponsors boycott the show, and some advertisers and about forty TV stations complied. But despite the criticism, redheaded actor David Caruso, who played tough yet sensitive Detective Kelly, and who bared his own slightly puffy derriere in a locker room scene in an October, 1993, episode, became a new sex symbol.

Actor Dennis Franz, as cantankerous, overweight, and balding Detective Andy Sipowicz, also became an unlikely sex symbol when his character started a steamy affair with Assistant District Attorney Sylvia Costas (Sharon Lawrence), whom he initially courted with such niceties as "Ipso this you pissy little bitch," while grabbing his crotch. (Lawrence also bared her breasts on the show). The two were married in a Greek ceremony in a May 1995 episode.

Despite its title and plot, *NYPD Blue* is mostly filmed in Hollywood, showing footage of New York, with some location work done in the Big Apple as well.

The precinct house shown is really on Manhattan's Lower East Side, and is the same police building used on *Kojak*.

## What to See in Manhattan

▶ **The Ninth Precinct Building,** 321 East 5th Street (between First and Second Avenues). There is today no Fifteenth Precinct in Manhattan, as on the show, but this building was once the Fifteenth, and the number 15 is etched in the stone near the roof line.

▶ **The New York Supreme Court Building,** 60 Center Street (downtown). The show also filmed at this block-sized courthouse, and according to a court spokesman, Room 452 at the real courthouse was replicated on the *NYPD Blue* set in L.A.

## Fun Facts

• Actor Dennis Franz has made a career of playing cops, and Detective Sipowicz is his twenty-eighth police character. His cop parts include *Hill Street Blues* (in which he appeared in two roles as Sal Benedetto and later as Norman Buntz), and the short-lived spin-off show *Beverly Hills Buntz*. Franz also played a cop in the movie *Diehard 2*.

• Dennis Franz won an Emmy for his role as Sipowicz in 1994.

• David Caruso left the show in 1994 to pursue a movie career.

• Caruso's star replacement was Jimmy Smits, who earlier worked with cocreator Steven Bochco on *L.A. Law*, playing Hispanic lawyer Victor Sifuentes.

• Caruso earlier received great reviews for his appearance on *Crime Story*, although he was killed off in that show's pilot episode. He was also Richard Gere's bunkmate in the movie *An Officer and a Gentleman*.

• According to *Parade* magazine, the theme song for *NYPD Blue*, written by Mike Post, was designed to "capture the way New York City assaults the auditory senses." To get the proper tone, the magazine said, the songwriter "digitized the sounds of a grinding cheese grater and 1,000 grunting Japanese men stomping on a wooden floor."

• The first TV show to exhibit a nude male butt was *The Bay City Blues*, the short-lived baseball team drama, which was also produced by Steven Bochco.

• Amy Brenneman, who played Officer Janice Licalsi, majored in religion at Harvard, according to *TV Guide*.

## THE ODD COUPLE
**(September 1970–July 1975, ABC)**

Based on the Neil Simon play of the same name, *The Odd Couple* featured Jack Klugman as Oscar Madison, the messy roommate, and Tony Randall as Felix Unger, the very neat roommate in a classic comedy of mismatched partners. Both divorced, the two men share an apartment on Park Avenue in New York. (In the play, they live on Riverside Drive).

Felix is a former *Playboy* photographer, now a free-lancer. He and his wife, Gloria (Janis Hansen), lived in New Rochelle, until their breakup. Oscar is a sportswriter for the *New York Herald*. His ex-wife Blanche (Brett Somers) visits occasionally, but Oscar sees more of Murray (Al Molinaro), a cop friend, and his other poker pals.

*The Odd Couple* lived in this East Side building. *(Photo by E. L. Tobias.)*

The show was initially filmed with a canned laugh track, but beginning in 1971, it was done before a live studio audience. The change came after Klugman and Randall led a successful and unprecedented campaign to eliminate the laugh-track machine.

The actors were less successful in their pitch to shoot the show in New York. It was done in Hollywood, with some exteriors, including the Park Avenue apartment building seen in the intro, shot in Manhattan.

## What to See in Manhattan

▶ **The apartment building at 1049 Park Ave.** (between East 86th and East 87th streets). This Upper East Side apartment building was seen as itself on the show, with the real address used as that of Felix and Oscar's. There is no real apartment 1102, however.

## Fun Facts

- Both Jack Klugman and Tony Randall were friends of playwright Neil Simon's brother, Danny, a TV writer whose divorce gave Neil Simon the idea for the show. Danny is said to be the inspiration for Felix.
- In real life Jack Klugman was married to Brett Somers, who played Blanche Madison, character Oscar's wife, but they divorced in 1974.
- On the show, Felix (Tony Randall) eventually got back together with his ex-wife Gloria (Janis Hansen).
- Garry Marshall *(Happy Days)* created the series. His sister, Penny, played Oscar's secretary, Myrna Turner. She later costarred in Garry Marshalls' *Laverne & Shirley.*
- In 1982, ABC introduced *The New Odd Couple,* a black version of the story, with Ron Glass (*Barney Miller)* as Felix, and Demond Wilson *(Sanford and Son)* as Oscar. The follow-up show used some of the same scripts as the original. It was not a hit.
- A cartoon version of the show, *The Oddball Couple,* featured a pairing of a neat cat and a messy dog.
- The 1968 movie version of *The Odd Couple,* also based on the hit Broadway play, starred Walter Matthau and Jack Lemmon.
- Actor Jack Klugman underwent surgery on his vocal cords in 1989, as a result of throat cancer, but was able to build his voice back. He and Tony Randall did a TV special of *The Odd Couple* in 1993, and then hit the road performing the Neil Simon play in theaters across the country.

## SEINFELD
**(May, 1990–     , NBC)**

The masturbation episode of *Seinfeld* was one of the funniest things ever on TV. Well, at least that's what the show's devoted fans will tell you.

Comedian Jerry Seinfeld, with his uncanny ability to put *everything* under a comedic microscope, has gained a cultlike following with this sitcom about love and life among thirtysomething singles in Manhattan.

A good week on *Seinfeld* provides coffee-time chatter for days, like the low-fat yogurt that was too good to be true episode, or the one in which Jerry tried to guess the name of an old girlfriend with the clue that her name rhymes with a body part, or the bubble boy episode, or the time lovable loser George (Jason Alexander) converted to Latvian Orthodox to get a date.

And then there are the fans who have started to see Seinfeldisms in their daily lives. "Did you ever wonder . . ."

As neurotic/New York as the dialogue on the show is, like most sitcoms, *Seinfeld* is filmed in Hollywood (some lucky fans even get into the tapings).

And the real joke may be that the apartment building seen on the show as Seinfeld's quintessential New York digs is not, in fact, on West 81st Street, as on the show, but really in L.A. According to the *FanFare* newspaper, clever fans have noticed the apartment building has earthquake reinforcement plates across the top, a sure giveaway it's on the West Coast.

Seinfeld and other cast members reportedly frequent an L.A. deli, but the exterior of the coffee shop on the show, at least, is truly New York.

The comedian repeatedly tells interviewers he really is like what viewers see on the show.

### What to See in Manhattan

▶ **Tom's Restaurant,** 112th Street and Broadway. This upper Manhattan coffee shop has become an increasingly popular tourist attraction since its appearance as the coffee shop on *Seinfeld.* It was earlier pretty much a Columbia University student and neighborhood hangout. The restaurant was also the subject of the Suzanne Vega song "Tom's Diner." The inside is not duplica-

*Seinfeld* has made Tom's Restaurant, on the upper West Side, a landmark. *(Photo by E. L. Tobias.)*

ted exactly on the *Seinfeld* set. Open weekdays 6 A.M. to 1:30 A.M., and twenty-four hours a day on weekends. Call 212-864-6137.

▶ **Royale Pastry Shop,** 237 West 72nd Street (between Broadway and West End Avenue). This Kosher bakery became popular with *Seinfeld* fans after the outside was seen on the show as the source for babka, a European coffee cake. Chocolate was the featured flavor on the show, but cinnamon is also sold here. Also popular with *Seinfeld* fans are the black and white cookies (with chocolate and vanilla frosting) mentioned on the show. The cakes sell for about $8, but are frequently on special. The cookies go for $1 apiece. Call 212-874-5642.

## Fun Facts

• Most of geeky Kramer's (Michael Richards) loud shirts are found in second-hand stores. The actor's hair is real. Richards parlayed his success on the show into commercial success, doing TV spots for Pepsi, paired with none other than Supermodel Cindy Crawford.

- Jason Alexander had a supporting role in the Ron Howard-directed movie *The Paper*, as a really pissed off parking commissioner.
- Playing the dad of George (Jason Alexander) on the show is Jerry Stiller, the real-life dad of Ben Stiller, director of the movie *Reality Bites* and star of the short-lived *The Ben Stiller Show*, which aired on Fox.
- Julia Louis-Dreyfus was earlier on *Saturday Night Live*, and does commercials for Clairol in her spare time. Unlike character Elaine, Louis-Dreyfus is married (to another former SNL alum, Brad Hall), and has a son.
- In addition to creating, producing, and starring in his hit show as himself, Brooklyn-born Jerry Seinfeld wrote a best-selling book, *SeinLanguage*, and has become a pitchman for American Express.
- Jerry Seinfeld told *Entertainment Weekly* that the best part of being rich is that he is able to "buy toothpaste in bulk."
- The show's pilot (in 1989) was called *The Seinfeld Chronicles*.

## TAXI
**(September 1978–July 1983, ABC, then NBC)**

The ensemble cast of this well-written sitcom included several wannabe characters working part-time as New York cabbies.

Tony (Tony Danza) wants to be a champion boxer. Bobby (Jeff Conaway) wants to be an actor. And Elaine (Marilu Henner) wants to work in the art world.

They are joined together by their jobs driving cabs for the Sunshine Cab Company in Manhattan, where Alex (Judd Hirsch), a full-time cabby, is on hand to offer fatherly advice.

The show's other characters are a colorful bunch, and include Louie De Palma (Danny DeVito), a short, nasty, money-hungry dispatcher who works in "the Cage"; Latka Gravas (Andy Kaufman), an immigrant of unknown origin who speaks English with

This one-time taxi garage was featured as such on *Taxi*. *(Photo by E. L. Tobias.)*

an hilariously broken accent; Simka Gravas (Carol Kane), a wacky native of Latka's home country who marries Latka in 1981; and Reverend Jim (Christopher Lloyd), a drugged out preppy who was ordained by the Church of the Peaceful.

The show was a hit from 1978 to 1980, but then ratings declined, and in 1982, ABC canceled *Taxi*. It was brought back by NBC, but just for another year.

The show was filmed in Hollywood, but footage of Manhattan was shown in its intro, including a real New York cab garage. And for establishing shots, cabs were shown driving on the real streets of New York and over recognizable area bridges, including the Queensboro Bridge.

## What to See in Manhattan

▶ **Dover Garage,** 534 Hudson (in the West Village). This garage once housed cabs and doubled on the show for the fictional Sunshine Cab Company. In 1989, the building was converted to a parking garage and gas station.

### Fun Facts

- After playing Louie, Danny DeVito hit it big in movies, and his wife, Rhea Perlman, hit it big on *Cheers*. Perlman appeared on *Taxi* as Zena Sherman, a candy-vending-machine person who briefly dates Louie.
- Andy Kaufman received great reviews for his bits, including Elvis, on *Saturday Night Live*, but his career was cut short by lung cancer. He died in May 1984, at age thirty-six.
- Marilu Henner's off-screen loves included Tony Danza, Judd Hirsch and John Travolta *(Welcome Back, Kotter)*. Henner later played Burt Reynolds's wife on *Evening Shade*.
- Christopher Lloyd went on to play Doc in the *Back to the Future* movies, with Michael J. Fox *(Family Ties)*.
- Tony Danza went on to the hit series *Who's the Boss*.
- Actress Carol Kane won two Emmys for her role as Simka Gravas, wife of Latka (Andy Kaufman) on *Taxi*.
- Judd Hirsch later starred in *Dear John*. He won Emmys for his role as Alex on *Taxi* in 1981 and 1983.
- The show was created by four former writers for *The Mary Tyler Moore Show*. And it was directed by James Burrows, who later helped create *Cheers*.

## FYI: ALSO IN NEW YORK CITY

The quirky New York–based, half-hour show *The Days and Nights of Molly Dodd* (May 1987 to August 1987, and March 1988 to June 1988, NBC, then on Lifetime from 1989 to 1991) was loved by critics, who coined a new phrase by calling it "dramedy," a combination of drama and comedy.

The show, which starred Blair Brown as Molly Dodd, a divorced woman in her mid-thirties, showed some New York sights. But most scenes were filmed inside a New York studio.

*The Days and Nights of Molly Dodd* featured this West Side diner. *(Photo by E. L. Tobias.)*

## What to See in Manhattan

- **The Empire Diner,** 210 Tenth Avenue (at 22nd Street). This twenty-four hour diner appeared on the show, and has also been featured in numerous movies and TV commercials. Specialties of the house include the turkey platters, and mashed potatoes with gravy. Call 212-243-2736.

# NEW YORK CITY ENVIRONS

New York City's boroughs are mostly residential areas, often ignored by tourists, but home to real New Yorkers, many with the fabulous one-of-a-kind accents that TV has tried so hard to imitate.

Make no bones about it, Archie Bunker was a Queens kind of guy, and the house seen as Archie and Edith Bunker's on *All in the Family* is in a Queens neighborhood that will seem familiar to fans of the sitcom.

Buddies Ralph and Norton were proud to be from Brooklyn *(The Honeymooners)*, as were the Sweathogs *(Welcome Back, Kotter)*. Both shows took inspiration from the real-life settings in which their stars were raised.

"A girl can only see the sights a girl can see from Brooklyn Heights," and that historic neighborhood was home to Patty and Cathy Lane on *The Patty Duke Show* (with the setting of the sitcom based on real-life Remsen Street in Brooklyn Heights).

The Huxtables *(The Cosby Show)* also lived in an upscale Brooklyn neighborhood (although the house shown on the show as their home is really in Manhattan).

And the Bensonhurst area of Brooklyn, in the 1950s, was featured on *Brooklyn Bridge*. While the setting seen was mostly a re-creation on a Hollywood set, shots of the real Brooklyn Bridge, which connects lower Manhattan and Brooklyn, were also shown.

Where would Officers Toody and Muldoon be if not in the Bronx *(Car 54, Where Are You?)*? That comedy show featured real shots of New York, but the Fifty-third Precinct was unfortunately make believe.

Outside the city, the suburb of New Rochelle is immortalized as Rob and Laura's address on *The Dick Van Dyke Show.* It's also where the show's creator, Carl Reiner, once lived.

The Seaver family on *Growing Pains* lived on Long Island; and Ann Marie *(That Girl)* hailed from the village of Brewster. But the real locations weren't featured on those shows.

## ALL IN THE FAMILY
**(January 1971–September 1983, CBS)**

When the first episode of *All in the Family* aired, an announcement preceded the show that warned the program "seeks to throw a humorous spotlight on our frailties, prejudices, and concerns. By making them a source of laughter we hope to show, in a mature fashion, just how absurd they are."

Bigoted Archie Bunker (Carroll O'Connor), star of this highly popular sitcom, and his family, including wife Edith (Jean Stapleton), lived at 704 Hauser Street in Corona, Queens. The fictional Hauser Street name on the show came from a street name in L.A.

In real life, however, the row of houses shown as Archie's neighborhood at the beginning of each episode is the 89–66 to 89–88 block of Cooper Avenue, in the Glendale section of Queens.

The real-life locale is an average, working class, mostly white neighborhood. Visitors will easily be able to visualize Archie living here.

Interiors for the show were done in Hollywood, where *All in the Family* was the first sitcom to be videotaped before a live audience.

### WHAT TO SEE IN THE GLENDALE SECTION OF QUEENS, NEW YORK

▶ **The house at 89–70 Cooper Avenue.** This private residence is seen as Archie's home on the show. Note the graveyard, unseen on the show, across the street.

The house seen as the home of the Bunkers on *All in the Family*. *(Photo by Fran Golden.)*

## Fun Facts

• The show was originally to be called *Those Were the Days*. ABC paid for the pilot, but rejected the show, which CBS then signed.

• Actress Jean Stapleton spent years trying to prove she is really nothing like the character she portrays. Stapleton refused to do the Edith character in interviews and once told a *Daily News* reporter, "I'm a one-woman committee to inform the public that playing a character is what acting is all about."

• Sally Struthers (Gloria), Carroll O'Conner (Archie), and Jean Stapleton (Edith) won Emmys for their performances on the show for the 1971–1972 season. Rob Reiner (Mike) would have to wait until the 1973–1974 season for his.

• Series creator Norman Lear was quoted as saying he patterned Archie after his own father, a salesman in Hartford, Connecticut.

• In 1972, the Teamsters Union complained about references in the press to "the Archie Bunker vote," saying it felt that working class people were being portrayed as "stupid."

• *All in the Family* was based on *Till Death Do Us Part,* a hit British show about a working class family, which also starred a bigoted father character with a live-in son-in-law.

• According to a 1974 article in *Parade* magazine, *All in the Family* itself generated a spin-off in Germany, *Ein Herz und Eine Seele* (A Heart and a Soul), whose main character, Alfred Tetzlaff is "anti-everything," including young women in miniskirts.

• Spin-offs of the show included *Maude* (she's Edith's cousin) and *The Jeffersons* (neighbors of the Bunkers until they move to posher quarters in Manhattan).

• The theme song, "Those Were the Days" was released as a single and as part of an LP called *All In The Family,* which rose to the top ten on the record charts. O'Connor cowrote the show's closing theme song, "Remembering You," with Roger Kellaway.

• Carroll O'Connor himself grew up in Queens, but is known to be more of a liberal thinker than Archie. He opposed the Vietnam War, and worked for presidential candidate Eugene McCarthy's campaign.

• In real life, Rob Reiner married actress Penny Marshall *(Laverne & Shirley),* and just like Mike and Gloria on *All in the Family,* they served Chinese food at their wedding. Reiner and Marshall were later divorced.

• Henry Fonda hosted the one-hundredth anniversary broadcast special of *All in the Family* in 1974. The two-hundredth episode anniversary show was hosted by Norman Lear in 1979.

• President Jimmy Carter played host to Carroll O'Connor and Jean Stapleton at the White House in 1978, the year their Archie and Edith chairs were donated to the Smithsonian. After the donation was made, it was decided to renew the show, as

*Archie Bunker's Place*, and the chairs, which had been purchased originally from a second-hand store, were reproduced at a cost of thousands of dollars.

• In one episode, Edith got away from a rapist by screaming "My cake is burning," and slamming the cake in the rapist's face.

• Character Edith died of a stroke on the show in 1980. That same year, Norman Lear and Tandem Productions donated half a million dollars to the National Organization for Women's Legal Defense and Education Fund, to establish the Edith Bunker Memorial Fund. The fund was dedicated to the ratification of the Equal Rights Amendment and women's rights causes.

• In 1994, Norman Lear developed *704 Hauser Street*, starring the Cumberbatches, a black family living in Archie Bunker's former home. The original door and a replica of the *All in the Family* set were used on the new show.

• Rob Reiner is the son of classic television funnyman Carl Reiner (who among other things created *The Dick Van Dyke Show* and played Alan Brady on that show).

## Fun Feature

Sammy Davis, Jr., guest-starred in one episode despite the show's policy of not hiring big-name performers. Davis said he spent months trying to persuade Norman Lear, the show's producer, to let him appear on the show, because he was such a devoted fan.

In the February 19, 1972, episode, Davis leaves his briefcase in Archie's cab and after calling the cab company comes to Archie's house to pick it up.

Impressed by the presence of the star, Archie tries to play the perfect host, but he manages to inadvertently mention Davis's glass eye and to offend Davis with numerous racial blunders. Davis responds politely, but when he and Archie pose for a photo, Davis shocks bigoted Archie by giving him a big wet kiss on the cheek.

The singer had been known to change his night-club performance schedule so he would be able to watch *All in the Family* on Saturday nights, and even plugged the show in his stage acts.

Davis said he was "speechless" when the offer to appear on the show came, and so excited the day of the taping that he woke up at 7 A.M. for an 11 A.M. call. "My wife thought I was ill. I'm a late sleeper and not known for being prompt, but on this special occasion I was fifteen minutes early. I couldn't wait," Davis said.

**ARCHIE-ISMS**

Meathead (Mike)
Dingbat (Edith)
Fag Country (England)
That Dago artist (Michelangelo)
A Dreaded Disease (Women's Lib)
Dumb Polack (Mike)
Little Girl (Gloria)
Stifle (Be quiet)
Yids (Jews)
Spades (Afro-Americans)

## THE HONEYMOONERS
**(October 1955–September 1956, CBS)**

Two of televisions's most popular comedy characters ever are loud-mouth Ralph Kramden (Jackie Gleason) and his best pal, Ed Norton (Art Carney), a bus driver and a sewer worker, who live in a tenement in Brooklyn.

The surroundings were sparse, Kramden sharing a two-room flat with his wife, Alice (Audrey Meadows, then Sheila MacRae), and Norton (rarely was he called by his first name) on the floor above with his wife, Trixie (Joyce Randolph, then Jane Kean).

On the show, the Kramdens and Nortons lived at 328 Chauncey Street, in the Bensonhurst section of Brooklyn, and that's the same address where Gleason himself was raised, except that the real street is in the Bushwick section of Brooklyn.

*The Honeymooners* was a series for only the 1955–1956 season, and thirty-nine episodes were filmed. The characters appeared for years, however, on Gleason's variety show, *The Jackie Gleason Show.*

Star Gleason reportedly refused to rehearse *The Honeymooners* skits, which were filmed before a live studio audience in New York (and later Miami). If he forgot a line or lost his place in the script, Gleason would rub his belly as a signal to his fellow players to improvise.

## What to See in Brooklyn, New York

▶ **The apartment building at 328 Chauncey Street.** This is where Jackie Gleason was raised and where the Kramdens and the Nortons live on the show. The building was not shown on *The*

The apartment building where Jackie Gleason was raised, and the address mentioned on *The Honeymooners.* *(Photo by Fran Golden.)*

*Honeymooners,* however. The neighborhood is not advisable for tourists.

## Fun Facts

- Pert Kelton, who played Alice on *Cavalcade of Stars* on the DuMont Network, the show on which *The Honeymooners* was first introduced, later played Alice's mother, whom Ralph referred to as Blabbermouth.
- WPIX-TV, Channel 11 in New York, has shown reruns of *The Honeymooners* on the air, late at night, for more than thirty years.
- Ralph and Norton belong to the International Order of Friendly Raccoons. Ralph is treasurer and Norton is the sergeant-at-arms. They also play together on the Hurricanes bowling team.
- Ralph's tone to Alice was sometimes threatening, as in "To the moon, Alice. One of these days, Pow, right in the kisser." But he would always end squabbles by telling his wife she is "the greatest."
- Jackie Gleason, who died in 1987, is buried in a mausoleum at Our Lady of Mercy Cemetery in Miami. "And away we go!" his popular phrase from *The Jackie Gleason Show,* is chiseled on the steps.
- In addition to Ralph Kramden, comedian Gleason's characters included The Poor Soul, Joe the Bartender, The Loudmouth, and Reggie Van Gleason III.
- Jackie Gleason tried to do a quiz show in 1961, called *You're in the Picture,* but it failed immediately.
- In the 1980s, "lost episodes" of *The Honeymooners* were found. Actually, they were sketches that had been on *The Jackie Gleason Show* and were cleverly pieced together to make new episodes.
- Ralph initially earns $42.50 a week for driving his bus route, which is eventually raised on the show to $62.
- Jackie Gleason wore around size 54 for much of his career, and earned the moniker The Great One.
- Ralph drives a bus on the Madison Avenue route for the Gotham Bus Company.

## WELCOME BACK, KOTTER
**(September 1975–August 1979, ABC)**

On this high school sitcom, the school was James Buchanan High School in Brooklyn. But the show was based on Comedian Gabriel Kaplan's own real-life experiences at Erasmus Hall High School in Brooklyn, where he had been an underachieving student.

On the show, Kaplan plays Gabe Kotter, an in-the-know teacher of inner-city remedial high school students. His charges include the Sweathogs, and the students are frequently invited to Kotter's home to talk about life.

John Travolta, as the tough leader of the gang (tame by today's standards), got a lot of attention for his performance as Vinnie Barbarino, and during the show's run became a superstar with his lead in the quintessential seventies film *Saturday Night Fever.*

The other Sweathogs included Juan Epstein (Robert Hegyes), Arnold Horshack (Ron Palillo), and "Boom Boom" Washington (Lawrence Hilton-Jacobs).

Travolta, Kaplan, and the show's producer, James Komack, were all high school dropouts.

The show was filmed at ABC-TV Center in L.A.

### WHAT TO SEE IN THE FLATBUSH SECTION OF BROOKLYN, NEW YORK

▶ **Erasmus Hall High School,** 911 Flatbush Avenue (at Church Avenue). This city high school has a student population of about 2,500.

### FUN FACTS

- The show was banned in Boston, where busing was a hot issue and it was decided by the ABC affiliate that rebellious students at an integrated school were not something to joke about. *Space: 1999* was shown instead.
- Only one of the actors playing a member of the Sweathogs, Lawrence Hilton-Jacobs, as "Boom Boom" Washington, was actually from New York City.

• On the show, Kotter's wife, Julie (Marcia Strassman), gave birth to twins in 1977.

• John Sebastian wrote and sang the theme song, "Welcome Back."

## FYI: ALSO IN NEW YORK ENVIRONS

While *The Dick Van Dyke Show* (October 1961 to September 1966, CBS) takes place in the suburb of New Rochelle, New York, the classic comedy, which starred Dick Van Dyke and Mary Tyler Moore as Rob and Laura Petrie, was shot in a studio.

Writer/director/actor Carl Reiner, who created the show, really did live in New Rochelle, however. And Reiner based the Petries—Rob, a TV comedy writer, and Laura, a housewife—on his real family.

On the show, Reiner himself played arrogant TV host Alan Brady, Rob's boss.

### WHAT TO SEE IN NEW ROCHELLE, NEW YORK

▶ **The private house** at 48 Bonnie Meadow Road (off Grand Boulevard). This was Carl Reiner's residence, when he worked as a comedy writer and performer in New York, according to local sources. On the show, the Petries also live on Bonnie Meadow Road, but at No. 148.

# BALTIMORE, PHILADELPHIA, AND WASHINGTON, D.C.

Movie director Barry Levinson made a name for himself doing movies based in his hometown of Baltimore, including *Diner*, so it is not surprising that he also turned to the city of his youth for *Homicide: Life on the Street.* The realistic-looking police drama, of which Levinson is an executive producer, is filmed entirely on location in Baltimore, and features shots of historic Fell's Point and other areas of the city.

That the plot of *Amen* was based in Philadelphia was also not a coincidence. That series's star, Sherman Hemsley, hails from the City of Brotherly Love. And a real West Philadelphia church was seen as a setting on the religious sitcom.

*thirtysomething* took place in Philadelphia too, but the show was shot in L.A., and it was a house in South Pasadena, California, that appeared as the Philadelphia home of Michael and Hope.

Washington, D.C., is seen in numerous shows, both as a setting for the plot and as a place characters visit as tourists.

Maxwell Smart *(Get Smart)* and Murphy Brown *(Murphy Brown)* are both based in the nation's capital. Inspector Lewis Erskine *(The F.B.I.)*, also calls the city home when he is not off tracking the F.B.I.'s Most Wanted. And Hawk *(A Man Called Hawk)* moved to Washington to keep the city's streets safe.

Among visitors to the city have been *The Beverly Hillbillies*, who came to town for a week, as tourists, in 1970, to see the

Lincoln Memorial, the Washington Monument, the Capitol, Lafayette Square, the Pentagon, the Smithsonian Institution's National Zoological Park, the Supreme Court Building, and Dulles International Airport (all shown on the show in exterior shots).

Other shows featuring Washington, D.C., as a setting included *Scarecrow and Mrs. King* and the short-lived, Washington-based *Lime Street* and *Capital News.*

Also of interest to TV fans, the Smithsonian Institution has among its massive collections real sets from *M*A*S*H,* Fonzie's leather jacket from *Happy Days*, and Archie and Edith's original chairs from *All in the Family.*

## AMEN
**(September 1986–July 1991, NBC)**

Philadelphia native Sherman Hemsley *(The Jeffersons)* suggested to series creator Ed Weinberger (cocreator of *The Cosby Show* and *Taxi*) that this religious sitcom be based in the City of Brotherly Love.

Hemsley, who played feisty Deacon Ernest Frye on the show, told *The Philadelphia Inquirer* that he advised Weinberger there is "a lot of churchgoing in Philadelphia," so basing the show here made sense.

In search of inspiration, the producers looked at photos of a number of real churches in the city, choosing the gray granite Mount Pisgah African Methodist Episcopal Church to represent the show's fictional First Community Church.

The real-life church's exterior was seen in the *Amen* intro, and the show also filmed some interior shots of Hemsley walking down the aisle of Mount Pisgah's sanctuary. The church's pastor, Reverend James E. Dandridge, did not allow the actors to preach from his pulpit, however.

Most of the show's interior scenes were taped in Hollywood, and the interior of the First Community Church is not really that similar to the real Mount Pisgah.

## What to See in Philadelphia, Pennsylvania

▶ **The Mount Pisgah African Methodist Episcopal Church,** 41st at Spring Garden Street (in West Philadelphia). The church, which appeared on the show, was organized in 1833, and moved to its present location in 1942. Sunday services are at 11 A.M., with Sunday school at 9:30 A.M. All are welcome to attend.

## Fun Facts

• The show was produced by Carson Productions Group Ltd., the production company owned by *The Tonight Show*'s Johnny Carson.

• Sherman Hemsley, who grew up in South Philadelphia, was reportedly a gang member in his youth and later worked as a post office clerk to support his early acting career. His debut on Broadway was in the hit musical *Purlie* (with Melba Moore).

• Guest stars on *Amen* included Emmy award-winning actress Jackee *(227)*, and the Godfather of Soul, James Brown, who sang "I Feel Good" in the show's final episode. M.C. Hammer also made his TV acting debut on *Amen.*

• While Sherman Hemsley is loud and cranky on *All in the Family, The Jeffersons*, and *Amen*, he is not known to be so in real life.

Costarring on the show was Clifton Davis *(That's My Mama)*, as Reverend Reuben Gregory. Davis is, in real-life, a Seventh-Day Adventist minister.

# THE F.B.I.
**(September 1965–September 1974, ABC)**

The cases of Inspector Lewis Erskine (Efrem Zimbalist, Jr.), the business-suited F.B.I. agent in this long-running crime show, were composites of real incidents in actual F.B.I. cases, or were based on what could have happened in real cases, according to Larry Hein, a retired F.B.I. agent.

*The F.B.I.* was endorsed by J. Edgar Hoover, the long-time

director of the real Federal Bureau of Investigation, and agents like Hein were assigned to make sure it was realistic.

"The reason the show lasted nine years is that we insisted on authenticity and absolutely no unnecessary violence," Hein said. "There were few people shot and killed on the show. J. Edgar Hoover was adamant there be no unnecessary violence."

Hoover apparently liked what he saw, and *The F.B.I.* crew was allowed to film inside the Department of Justice building (where the F.B.I. was based until moving into its own building in 1974), including at the agency's crime laboratory.

*The F.B.I.* also showed the outside of the Justice Department building, and other Washington sites such as the U.S. Capitol, the Washington Monument, and the Jefferson Memorial, to establish locale.

The show was really shot on a Hollywood set, but the set was a very authentic re-creation of the agency's Justice Department digs, according to those in the know.

At least once a year, actor Zimbalist would be filmed in Washington, with a prototype of the newest car of the Ford Motor Company. Ford was a sponsor of the series and Zimbalist would ride the company's new model out of the Justice Department building's 10th Street entrance.

At the end of some episodes, the real F.B.I. used the forum to ask viewers for information on its most wanted criminals. A trailblazer for the later *America's Most Wanted* show, *The F.B.I.* would display a picture of the fugitive with a description.

Agent Hein said some fugitives were actually captured based on viewer responses.

## What to See in Washington, D.C.

- **The Department of Justice Building,** 9th and Pennsylvania Avenue NW. This was the building that was seen on the show, and where the F.B.I. was based until moving to its own headquarters nearby. The building is not open to the public.

- **The Federal Bureau of Investigation Headquarters,** 10th and Pennsylvania Avenue NW. Visitors can take a free one-hour tour, which includes a walk past the agency's real crime laboratories, a live firearms demonstration, and a history lesson on the

The headquarters of the Federal Bureau of Investigation. *(Photo courtesy of the U.S. Department of Justice, Federal Bureau of Investigation.)*

FBI. Tours are offered Monday to Friday, 8:45 A.M. to 4:15 P.M., except on federal holidays. Call 202-324-3447.

## FUN FACTS

• Efrem Zimbalist, Jr., is the father of Stephanie Zimbalist. On her show, *Remington Steele*, he played a criminal, not a crime fighter.

• Actor Zimbalist earlier starred as Stu Bailey, the highly educated private detective on *77 Sunset Strip.*

• In real life, Zimbalist's parents were famous musicians, a violinist and an opera singer.

• A remake of *The F.B.I.* in 1981, called *Today's F.B.I.*, starred Mike Connors *(Mannix)*, but was not a hit, and lasted less than a year.

## GET SMART
**(September 1965–September 1970, NBC, then CBS)**

"Would you believe . . ." Comedian Don Adams won three Emmy awards for his portrayal of bungling secret agent Maxwell (Max) Smart, Agent 86, on this comedy spy show.

Max, who works for the fictional government agency CONTROL, in Washington, D.C., speaks several languages and knows karate and judo, but manages to mess up regularly, despite such spy accoutrements as shoe phones, in his efforts to battle enemy KAOS. His boss, The Chief (Edward Platt), frequently hears from Agent 86, "Sorry about that, Chief."

Max eventually falls in love with CONTROL's beautiful Agent 99 (Barbara Feldon), who patiently roles her eyes at him when he makes mistakes, "Oh Max." Although he never learns her real name, Max and 99 marry and have twins, a boy and a girl.

The show's intro showed Max jumping out of his blue convertible, going into the CONTROL headquarters building, passing through several sliding metal doors, and entering a phone booth with a bottom that opens, presumably dropping him into the top secret underground spy center.

Real shots of Washington, D.C., were featured in the intro in later episodes, to establish locale, including the Capitol, the Lincoln Memorial, the Jefferson Memorial, and the White House.

### What to See in Washington, D.C.

▶ **U.S. Capitol,** National Mall. The Capitol building is one of the sites seen in the show's intro, with the shot including the National Mall, a nearly two-mile-long green park, which runs between the Capitol and the Washington Monument. Free tours of the building are offered daily, 9 A.M. to 3:45 P.M. Visitors can also explore on a self-guided basis from 9 A.M. to 4:30 P.M. (and until 8 P.M. in the summer). Visitors to the Capitol should contact their local senator or congressman in advance for a pass to watch Congress in session. The building is closed Thanksgiving, Christmas, and New Year's Day. For tour information call 202-225-6827.

The U.S. Capitol. *(Photo courtesy of the Washington D.C. Convention and Visitors Association.)*

## Fun Facts

- Actor Robert Karvelas, who played Agent Larrabee, is Don Adams' real first cousin. In a lecture for the Museum of Broadcasting, Adams said Karvelas was willing to do dumb jokes that even Adams wasn't willing to do.
- Don Adams also performed the voice of two famous cartoon characters, both with their own shows, Tennessee Tuxedo and Inspector Gadget.
- Don Rickles was among the comedians who guest-starred on *Get Smart*, and Don Adams said working with Rickles was hard because they kept cracking each other up so much it was difficult to get through the scenes.
- Carol Burnett also guest-starred as a country-western performer who by mistake swallows an olive containing a miniature receiver.
- The show was created by veteran comedy writers Mel Brooks (also a noted movie director) and Buck Henry.

• Don Adams did his own fight sequences and once accidently got punched hard and broke his nose. He appeared in the subsequent episodes with a real cast.

• Barbara Feldon was a top New York model and a winner on *The $64,000 Question.*

• Fang is Max's spy dog pal, whose code name is K-13. Hymie (Dick Gautier) is his robot friend, who used to work for KAOS, but was reprogrammed and now works for CONTROL.

• Max (Don Adams), Agent 99 (Barbara Feldon), and other members of the *Get Smart* cast were reunited in 1989 in a TV movie, *Get Smart, Again.*

• A new weekly version of *Get Smart* debuted on the Fox network in 1995.

## HOMICIDE: LIFE ON THE STREET
**(January 1993– , NBC)**

Style and substance attracted fans, and raves from critics too, to this cop show, which premiered after the Super Bowl postgame show.

The ensemble cast of experienced and recognizable actors plays cops with emotion, whose lives are affected by the murders they see on their jobs as homicide investigators, but who can still laugh once in a while.

The show's stark realism is achieved by filming on location in Baltimore, the hometown known so well by executive producer Barry Levinson. Most noted as a movie director, Levinson earlier featured the city in his movies *Diner, Tin Men,* and *Avalon.*

Much of the location footage is shot around Baltimore's Fell's Point Neighborhood. The historic area has within its three blocks a "great treasure trove of American architecture," according to Ed Kane, operator of The Water Taxi, which connects fifteen locations around the Inner Harbor to Fell's Point. About 330 of the buildings date from the late eighteenth century or early part of the nineteenth century.

Today the buildings house antique shops, restaurants, and especially a lot of watering holes, which is nothing new. Said Kane, "A late eighteenth-century seafarer would recognize Fell's Point today."

The Fell's Point area of Baltimore. *(Photo courtesy of The Baltimore Water Taxi.)*

## What to See in Baltimore, Maryland

- ▶ **Recreation Pier,** 1715 Thames Street. The show uses this building as its police station location. The building was once used to house the offices of the harbormaster and other port authorities, and had a playground on top, thus its name. Visitors can take pictures outside, but the main set inside is off-limits.

- ▶ **Jimmy's Restaurant,** Lancaster and Broadway. In this long-established city diner, which appears on the show, visitors might find Senator Barbara Mikulski (D-Maryland) at one table, and bikers at another. First Lady Hillary Rodham Clinton has stopped by on occasion too. The food is inexpensive and home-cooked, and the place is very popular. Open daily, 6 A.M. to 10 P.M. Call 410-327-3273.

- ▶ **The Waterfront Hotel,** 1710 Thames Street. Located across the street from the show's police station, and also appearing on the show, this restaurant/bar traces its roots back to the days when sailors came here seeking carnal pleasure. Today the bar is a major local meeting place. Open 11 A.M. to 2 A.M. Call 410-327-4886.

▶ **The Wharf Rat,** 801 South Ann Street. The show filmed people eating crabs at this bar, but they aren't really served here. The Wharf Rat is really a world-class watering hole with dozens of varieties of beer, served in a building that dates back to the eighteenth century. Open 11:30 A.M. to 1 A.M. Call 410-276-9034.

## Fun Facts

• The show was inspired by David Simon's critically acclaimed nonfiction book *Homicide: A Year on the Killing Streets*.

• Robin Williams guest-starred in the premiere episode in 1993, in a serious role, as a tourist traveling with his family; his wife is killed when they wander, by accident, into a bad neighborhood.

• Most of the cast members had earlier appeared in movies. Ned Beatty, who plays Stanley Bolander, won an Oscar for his performance in *Network*.

• Yaphet Kotto, who plays Lieutenant Al Giardello, was raised Jewish in Harlem, is now a practicing Hindu, and is related to a royal family in Cameroon, according to *People* magazine. He earlier appeared in the movie *Alien*.

• Actor Richard Belzer, who plays John Munch, is better known as a stand-up comic, but also appeared in *Bonfire of the Vanities*.

• Daniel Baldwin, who plays Beau Felton, was in the movie *Born on the Fourth of July*, and he's Alec's brother, too.

• Melissa Leo, who plays Kay Howard, was earlier on *All My Children*.

• Executive producer Barry Levinson won an Oscar for his direction of the movie *Rain Man*.

• Critics called *Homicide: Life on the Street* the heir to *Hill Street Blues*.

## MURPHY BROWN
**(November 1988–  , CBS)**

Dan Quayle's favorite single mom, Murphy Brown (Candice Bergen), is a tough-talking news reporter on *F.Y.I.*, a TV newsmagazine show based in Washington, D.C.

Actress Bergen is said to have modeled the role in part on her friend, real-life newswoman Dianne Sawyer. The show's cocreator, Diane English, is a former newswoman (for PBS).

Murphy is not perfect. She's a recovering alcoholic, she's perhaps a bit too aggressive on stories, she drives her beloved Porsche too fast, she sings Motown songs badly, and she is definitely hard to work with.

But that's not why Vice President Dan Quayle picked on her. Quayle did not like the character's decision to have a baby on her own. The vice president attacked Murphy Brown in a 1992 speech, saying she was "mocking the importance of fathers by bearing a child alone and calling it just another life-style choice."

In a brave bit for the TV history books, Murphy talked back to Quayle in September 1992, before 70 million TV viewers, declaring "Perhaps it's time for the vice president to . . . recognize that whether by choice or circumstance, families come in all shapes and sizes." Quayle later sent Avery, Murphy's son (named for her mother), a stuffed elephant, which the show's producers donated to a homeless shelter.

*Murphy Brown* is popular not only with the public at large, but also with real TV news people, and several have appeared on the show. Guest star Linda Ellerbee, the newswoman and commentator, jokingly likened the sitcom to a documentary.

Production for *Murphy Brown* takes place not in Washington, but before a studio audience in Burbank, with most outside shots done at the Warner Brothers Ranch.

To establish locale, however, the show features stock footage of familiar sites in Washington, including the Capitol, the Washington Monument, and the Lincoln Memorial.

In one episode, Murphy takes her baby, Avery, to an Easter egg hunt at the White House, really hoping to get an interview with President Clinton. She ends up accidently tripping the President, not exactly endearing herself to the nation's leader.

When *Murphy Brown* visits the White House, the president knows it. *(Photo courtesy of the Washington D.C. Convention and Visitors Association.)*

On the show, Murphy lives in historic Georgetown, in a brownstone, but if there is a real historic home in Washington that is shown as Murphy's, it is a well-kept secret.

## What to See in Washington, D.C.

- **The White House,** 1600 Pennsylvania Avenue. A self-guided tour is offered Tuesday through Sunday, 10 A.M. to noon, free of charge. Tickets are required from April to October, and are available at the ticket booth just south of the White House, beginning at 8 A.M. on those days. Early arrival is recommended. The tours are not offered on Thanksgiving, Christmas, or Independence Day, or when special functions are taking place in the building.

- **Georgetown** (between Rock Creek Park and Georgetown University, and the corner of Reservoir Road and Wisconsin Avenue, to the Potomac River). The area features historic homes, boutiques, and interesting restaurants and nightclubs, and attracts a crowd of students and the upwardly mobile. It was once a colonial port and was also part of the Underground Railroad system.

▶ **Washington Monument,** National Mall at 15th Street NW. This tall monument to the first U.S. president was dedicated in 1885. There's a free elevator ride offered to the top. It's open daily at 9 A.M. and closed at midnight, from April 1 to Labor Day, and closed at 5 P.M. other times of the year. The monument is closed on Christmas Day.

## Fun Facts

• Murphy's (Candice Bergen) rules for her infant son, when she's out of town, include "No parties, no girls, and no matter how much you cry, you can't borrow the Porsche."

• Avery belongs briefly to the "Little Bitty Buddy Play Group," but is kicked out after Murphy (Bergen) insults the other moms with her own rendition of "The Wheels on the Bus."

• The part of baby boy Avery is actually played by baby girls.

• Barry Manilow played himself on a 1993 episode. He was a birthday gift to Avery from Frank (Joe Regalbuto). Manilow is Avery's favorite singer.

• Corky's (Faith Ford) husband, Will (Scott Bryce) hits it big with his first novel, *The Little Dutch Boy.* His success goes to his head, however, and the two divorce.

• Jerry Corley, son of Pat Corley, who plays Phil, did a cameo as Phil's nephew in a 1993 episode.

• The Museum of Broadcasting in New York was the setting for a 1992 episode of the show. Murphy is there to pick up an award.

• Joe Regalbuto, who plays Frank, also directed some episodes of the show.

• *CBS This Morning* did a broadcast from the *Murphy Brown* set, during which Bergen discussed motherhood and the Dan Quayle controversy, and the other actors previewed the 1992 TV season.

• Real-life TV newswomen Katie Couric, Faith Daniels, Joan Lunden, Mary Alice Williams, and Paula Zahn attended Murphy's baby shower in a May 1992 episode. Corky (Ford) was the hostess.

• Connie Chung and Irving R. Levine also appeared on one episode of the show, as did Walter Cronkite, who helped roast F.Y.I. Anchor Jim Dial (Charles Kimbrough), saying he told Jim not to go into TV because he was too skinny.

• Veteran actress Colleen Dewhurst played Murphy's mom, Avery. Dewhurst passed away in 1991, and Murphy copes with her Mom's death on the show. Dewhurst posthumously won a second Emmy for her performance on *Murphy Brown*, only two days after her death.

• Darren McGavin plays Murphy's dad, and won an Emmy for the role in 1990.

• Aretha Franklin is Murphy's idol, and the Queen of Soul herself guest-starred on the show in 1991. Murphy, who can't sing very well, joined Franklin for a rendition of "Natural Woman."

• Grant Shaud, who plays Miles Silverberg, is not Jewish and reportedly doesn't really wear glasses or suits. He is said to be modeled after the show's cocreator and co–executive producer Joel Shukovsky, husband of Diane English.

• Candice Bergen first showed her comic and singing skills in the movie *Starting Over* with Burt Reynolds. The actress once really worked as a journalist, reportedly turning down a job on *60 Minutes* during that show's early years.

• In real life, Bergen is married to movie director Louis Malle (who guest stars as a director in one episode of the show) and has a daughter, Chloe. In her spare time, the actress does commercials for Sprint.

# THE SOUTH

Despite the fact that it's far from Hollywood, the South has been chosen as production base by several TV producers, drawn not only by the change in scenery but also by good ol' Southern charm and hospitality.

Covington, Georgia, is the setting seen on *In the Heat of the Night* (although the plot takes place in Mississippi). The drama was shot on location in and around the small city, and visitors will recognize a number of real locations, including Covington's historic courthouse.

The same city is also featured in early episodes of *The Dukes of Hazzard*, where it is supposed to be in fictional Hazzard County.

*I'll Fly Away* also found a home in the Peach State, shooting on location in and around Madison, Georgia. The show was supposed to take place in a fictional Georgia town.

Also using Georgia as a backdrop was the short-lived *Breaking Away*, filmed on location in Athens, Georgia (although the show's plot takes place in Bloomington, Indiana).

Atlanta was the setting for the action on *Matlock*, but the lawyer show never filmed more than a few exteriors in Georgia's capital city. The show was really shot in L.A. That is, until 1993, when production for *Matlock* moved down south, not to Atlanta, but to Wilmington, North Carolina.

Wilmington was also home for production of *The Young Indiana Jones Chronicles*, although the plot for the show takes place in a variety of other locations.

Residents of Mount Airy, North Carolina, readily point out similarities between their town and fictional Mayberry on *The*

*Andy Griffith Show.* Although the sitcom never actually filmed there, Mount Airy is the proud hometown of Andy Griffith. And Mayberry's fame is celebrated annually in Mount Airy with a "Mayberry Days" festival.

Little Rock, Arkansas was the real setting for most of the familiar exteriors seen on *Designing Women,* including the Sugarbakers' home, although the plot for the sitcom takes place in Atlanta (and most of the production was done in Hollywood).

Arkansas is also the location for exteriors seen on *Evening Shade.* There are really two towns called Evening Shade in the state, but only one appears on the show.

Schuyler, Virginia, was never actually seen on *The Waltons,* but the outside of the house where the show's creator, Earl Hamner, Jr., grew up (which was duplicated on the show's Hollywood set) can be viewed in the small Blue Ridge Mountain town. There's also a museum in Schuyler that pays tribute to Hamner and the town's place in TV history.

In Eastern Virginia, fans of the military sitcom *Major Dad* will want to visit the real Marine Corps base at Quantico, which appears in exterior shots as fictional Camp Hollister on the show.

## THE ANDY GRIFFITH SHOW
**(October 1960–September 1968, CBS)**

There's no question in the minds of people in Mount Airy, North Carolina, that their small city was the model for Mayberry, the fictional town of traditional values and good neighborliness on *The Andy Griffith Show.*

Andy Griffith, who played Sheriff Andy Taylor on the show, was raised in Mount Airy. And even though the actor has often said Mayberry is not a real place, folks here say there's no denying the similarities.

A lot of real streets in the city of Mount Airy are the same as those on the show, including Rockford Street, Haymore Street, Spring Street, and, of course, Main Street. Real shops are also mentioned, and lots of the names of real people in town are familiar to fans of the show, as well. You may very well bump into an Emmett in Mount Airy, for instance.

There was a suggestion once that the city of just over 7,100 rename itself Mayberry. The citizens of Mount Airy wouldn't go quite that far, but they aren't shy about their place in TV history either.

The city, which is located at the foot of the Blue Ridge Mountains, about thirty-five miles northwest of Winston-Salem, annually hosts Mayberry Days, which takes place the last weekend in September. The festival features down-home activities including pie-eating contests, bake sales, and bluegrass music performances. And the event attracts thousands of visitors.

*The Andy Griffith Show* was really filmed entirely in California, but several sites mentioned on the show can be found in the "loop" in downtown Mount Airy, which includes the city's original Main Street.

## What to See in Mount Airy, North Carolina

▶ **Floyd's City Barber Shop,** 129 North Main Street. Operated for more than forty years as Main Street Barber Shop, this establishment changed its name in the 1980s, in homage to Floyd

The real Floyd's City Barber Shop. *(Photo courtesy of The Gilmer-Smith Foundation.)*

Lawson's establishment on the show. This is an old-time barber shop where you can still get your neck shaved with a straight razor, if you so desire. The less brave can pose for pictures in one of the two barber chairs, or have a manicure if the manicurist is on duty. T-shirts and other souvenirs are for sale.

▶ **Snappy Lunch,** 125 North Main Street. Next door to Floyd's, this is one of several real-life Mount Airy eating establishments mentioned on the show (the now-closed Weenie-Burger was another). Snappy Lunch dates back to 1923 and is famous locally for its pork chop sandwich. It's also a good place to hear Griffith trivia from the locals. The restaurant is open for breakfast, beginning at 5:30 A.M., and lunch. A Mayberry photo collection decorates the walls.

▶ **The Andy Griffith Playhouse,** corner of Rockford Street and Graves Street. This building was previously the Rockford Street Elementary School, which Andy Griffith attended, and where

The Snappy Lunch in downtown Mount Airy. *(Photo courtesy of The Gilmer-Smith Foundation.)*

Mayor Stoner (really former Mount Airy mayor Maynard Beamer) and Barney impersonator (actor David Browning) at Mayberry Days Festival. *(Photo by Robert "Dozy" Caldwell, courtesy of The Gilmer-Smith Foundation.)*

he first performed. The 350-seat auditorium is today used for local theater and musical performances. The building is also home to **The Surry Arts Council,** which coordinates the town's Mayberry Days event and offers walking tours for fans of the show. Call 910-786-7998.

▶ **The top of Spring Street sign.** On *The Andy Griffith Show* a running gag is the knocking down of the stop sign at Spring Street. The real Spring Street stop sign is located across from the Andy Griffith Playhouse, where Spring Street ends on Rockford Street.

▶ **Rockford and Haymore Streets.** This is the school crossing mentioned on the show. It's also where real-life Andy crossed to go to Rockford Street Elementary School. Griffith lived at 711 Haymore.

▶ **Lamm's Drug Store,** Main Street between Franklin and Pine Streets. This shop, mentioned on the show, was closed in 1990, but the fixtures remain intact, and Lamm's is reopened during Mayberry Days.

▶ **Old City Hall,** Moore Avenue. The town bought a 1962 Ford Galaxie and restored the car to look like Sheriff Andy Taylor's squad car. And the replica can be viewed outside this building. The Old City Hall actually did house the city's police department, jail, and mayor's office until 1978, when a new city hall was built. The building, which dates from the early 1900s, is today used by private businesses, including Friendly Heating and Cooling, which graciously allows visitors to pass through its offices to view the old jail. Call 910-789-6453.

▶ **Mayberry Bed & Breakfast,** 329 West Pine Street. Two rooms are offered to guests for about $40 each a night in this house just around the corner from the Andy Griffith playhouse. Call 910-786-2045.

▶ **Mayberry Motor Inn,** North Highway 52 Bypass. The motel offers twenty-seven rooms (about $38 each per a night), with one bedroom decorated with actual furnishings from Frances Bavier's (Aunt Bea's) estate. Call 910-786-4109.

▶ **Aunt Bea's Barbecue,** Highway 52 North. A very casual barbecue and ice cream venue, Aunt Bea's also features banana pudding and peach cobbler.

▶ **WPAQ Radio** (740 AM) This is the local place to tune in for old-time bluegrass and traditional music.

▶ **Art Market,** 148 Main Street. About 150 local artists offer items here on a consignment basis, including a number of **The Andy Griffith Show**-related prints and high quality souvenirs. The shop is run under the auspices of The Surry Arts Council.

## Fun Facts

• In 1993 Andy Griffith told *USA Today* that he had planned originally to be the funny man on the show, but realized by the second episode that Barney (Don Knotts) should be funny and that he should play the straight guy. "Then each time somebody came on the show who was funny, we made them a regular. That's how the town of Mayberry came to life," Griffith said.

• According to the Associated Press, Andy Griffith once thought about entering the Moravian ministry.

• Frances Bavier, who played Aunt Bea, retired to Silver City, North Carolina, and became a bit of a recluse after the show ended. She died in 1989 at the age of eighty-six.

• In real life, in 1962 Don Knotts was made a full-fledged deputy sheriff of Morgantown, West Virginia, his hometown. The oath was administered by the town's sheriff in New York City, where Knotts was rehearsing for a television guest appearance.

• The fishing hole, Meyer's Lake, seen on the series was really filmed at Coldwater Canyon, outside of L.A.

• Ron Howard, who played Opie, was known as Ronny at the time of the show, and grew up to be Richie Cunningham on *Happy Days*. He later became a hot movie director (see *Happy Days*).

• Jim Nabors, who played Gomer Pyle, was discovered by Andy Griffith while performing in a nightclub in Santa Monica, California. Nabors left Mayberry for his own series, *Gomer Pyle, U.S.M.C.*, and later continued his singing career.

• On the show, Floyd (Howard McNear) had only one barber's chair. And Sheriff Barney (Don Knotts) had only one bullet in his gun.

• The show spawned a spin-off, *Mayberry R.F.D.*

• When Don Knotts (Barney) left the show after five years to pursue a movie career, a new deputy was introduced named Warren Ferguson (Jack Burns). On the show, Barney joined the North Carolina State Traffic Commission in Raleigh.

• *The Andy Griffith Show* Rerun Watchers Club has more than twenty thousand members, with chapters around the world including in Australia and Saudi Arabia.

## DESIGNING WOMEN
**(September 1986–May 1993, CBS)**

FOB's (friends of Bill Clinton's) Harry Thomason and Linda Bloodworth-Thomason produced *Designing Women*, a show populated by Southern females and snappy repartee. Their Arkansas connection showed up in several ways on the popular series.

While the location for the plot of *Designing Women* is Atlanta, the real house shown as the headquarters for the Sugarbakers' decorating firm is pure Little Rock, not Georgia Peach, and is in the same neighborhood, in fact, as Bill and Hillary Clinton's former abode.

The Arkansas Governor's Mansion also appeared in early episodes of the show (when you-know-who were in residence), as Suzanne Sugarbaker's (Delta Burke) house.

And while Suzanne and sister Julia (Dixie Carter) are from Georgia on the show, the character of Charlene (Jean Smart) is from Little Rock.

As a show, *Designing Women* was not shy about taking on hot topics.

During the confirmation hearings for Clarence Thomas's appointment to the Supreme Court, the characters on the show were seen watching the televised sessions, some in the Anita Hill camp (she accused Thomas of sexual harassment) and others in the Thomas camp. In the episode, Mary Jo (Annie Potts) responds to Senator Alan Simpson's (R-Wyo.) real-life comment about "sexual harassment crap" by saying ever so politely, "nice talk from a U.S. senator."

One actress on the show credited her appearance to President Clinton. Sheryl Lee Ralph, who played Etienne Toussant Bouvier, a show girl who marries Anthony in Las Vegas, hosted a campaign fund-raiser for Clinton, at which she met Harry Thomason. She took the opportunity to criticize Thomason for not having a regular black female cast member. And he hired Ralph, whose previous acting credits included appearing on Broadway in *Dreamgirls*.

## What to See in Little Rock, Arkansas

► **Villa Marre,** 1321 South Street (at 14th). The house seen as the Sugarbakers' on the show is really a museum, located about seven blocks from the Governor's Mansion, in a rehabilitated inner-city area. The house has ten main rooms, and was built in 1881 by Angelo Marre, an Italian emigrant, and his wife, Jennie Marre. The house was altered extensively by its second owner in 1910. The exterior design shows Italian influence, while the decor of the museum dates from 1850 to 1910. The house was the first Victorian in Little Rock to be rehabilitated (it was in

Villa Marre was seen as the Sugarbakers' house on *Designing Women.* *(Photo courtesy of the Quapaw Quarter Association.)*

major disrepair in the early 1960s). And it was donated to the Quapaw Quarter Association in 1979 by James W. Strawn, Jr. The museum is open Monday through Friday, 9 A.M. to 1 P.M., and Sunday 1 P.M. to 5 P.M. Admission is charged. Call 501-374-9979.

▶ **The Arkansas Governor's Mansion** 1800 Center Street (at 18th and Spring Street). The official state residence appears as Suzanne Sugarbaker's house on early episodes of *Designing Women.* Exterior shots only were taken. The Georgian Colonial-style house was completed in 1950. Visitors can tour some of the rooms inside, and see the collection of furnishings and artwork that includes a sixty-two piece silver service that was used aboard the battleship *Arkansas*, as well as paintings of the house's residents, including then-governor Bill Clinton. Tours of the home are offered on Tuesday and Thursday, on an appointment-only basis. Call 501-376-6884.

## Fun Facts

- Former beauty queen Suzanne Sugarbaker's (Delta Burke) beauty tips include spitting in your frosted eye shadow to enhance the color.
- In real life Delta Burke met Gerald McRaney (star of *Major Dad*) on the set of *Designing Women* and they were married. On the show, he played Dash Goff, one of three of Suzanne's ex-husbands.
- Delta Burke left *Designing Women* in 1989, accusing the producers of, among other things, hassling her about her weight gain. On the show, character Suzanne moves to Japan.
- Delta Burke and the Thomasons settled their differences, and Burke went on to star once again as Suzanne Sugarbaker in the 1995 Thomason comedy *Women of the House*. On the show, Sugarbaker (Burke) has inherited her late husband's seat in the House of Representatives, and moves to Washington, D.C.
- Other guest stars on *Designing Women* included Richard Gilliland, Jean Smart's real-life husband. Dixie Carter's husband, Hal Holbrook (of *Evening Shade*), and her two daughters, Mary Dixie and Ginna Dixie also appeared on the show.
- Alice Ghostley (Bernice Clifton) earlier played Esmeralda on *Bewitched*.
- Delta Burke and Dixie Carter and Linda Bloodworth-Thomason and Harry Thomason also worked together on the earlier show *Filthy Rich*, a Memphis-based takeoff on *Dallas*. The show lasted only one season.
- Annie Potts went on to make popcorn commercials and star in *Love and War*.
- Comedian David Steinberg was executive producer of *Designing Women* for the final season.
- The Donald's (as in Trump) Marla Maples guest-starred on *Designing Women* in October, 1991.
- On the show, the Sugarbakers' decorating firm is at 1521 Sycamore Street, Atlanta.

## THE DUKES OF HAZZARD
**(January 1979–August 1985, CBS)**

This popular action-packed series featured Southern cousins, Luke Duke (Tom Wopat) and Bo Duke (John Schneider), who drove fast cars and generally caused trouble, especially for farcical bad guy, Boss Hogg (Sorrell Booke), and his brother-in-law, the bumbling sheriff of Hazzard County, Roscoe Coltrane (James Best).

The Dukes' red 1969 Dodge Charger, the "General Lee," was constantly crashing in well-executed stunts. The cars were often destroyed, and the show went through some three hundred General Lee cars during its run.

Alan Facemire, who worked as a crew member on the show, said some of the cars were old painted taxi cabs; the only prerequisite was that they could get up to forty-five miles an hour. After the stunt the cars were usually sent to the junkyard. The special effects team, which included mechanics and demolition experts, also blew up a fair number of barns, Facemire said.

The setting for the action was a fictional town somewhere east of the Mississippi and south of Ohio. While much of the series was shot in a Burbank studio, location work for the intro and six or so early episodes was done in and around Covington, Georgia, the same small city that would later be the setting for *In the Heat of the Night.*

### What to See in Covington, Georgia

▶ **The Railroad Crossing at Elm Street** (near downtown). A stunt jump on this railroad track was seen in the opening sequences of the show.

### What to see in Oxford, Georgia

▶ **Oxford College,** 100 Hamill Street. The campus and area around this two-year college were the setting for some chase scenes that were used on the series.

## Fun Facts

- The show was a big hit in England.
- Country star Waylon Jennings played the off-screen narrator of the series, known as "The Balladeer." And the theme song, "Good Ol' Boys," which was sung by Jennings, became a hit song.
- The stars of the show, actors Tom Wopat and John Schneider, walked out in 1982, in a contract dispute, returning in 1983. During their absence, featured were actors Byron Cherry and Christopher Mayer, who played Duke cousins Coy and Vance.
- Toys, games, and T-shirts were created featuring *The Dukes of Hazzard* name.
- The show spawned a spin-off, *Enos*, starring Sonny Shroyer as Enos Strate, who had been a deputy sheriff on *The Dukes of Hazzard*.

## EVENING SHADE
**(September 1990–May 1994, CBS)**

This sitcom, about high school football coach Wood Newton (Burt Reynolds) and other characters in the sleepy small town of Evening Shade, Arkansas, was produced by the same team that did *Designing Women*, Linda Bloodworth-Thomason and Arkansan Harry Thomason (both friends of President Bill Clinton). And the show was the first on the air to feature Arkansas as the location for its plot.

The Thomasons said they chose the name *Evening Shade* for the series because "it has a wonderful sound" when pronounced.

There are really two towns in the state of Arkansas with the name Evening Shade. The one seen occasionally on the show is a town of four hundred, in Sharp County, which is in northern Arkansas. The other is in southwest Arkansas and is even smaller than the first, according to state tourism officials, who had a hard time finding it on the map.

According to local legend, Evening Shade (in Sharp County) got its name from large trees, which afforded refuge from the sun at the town's old grist mill.

Once a booming lumber town, Evening Shade has today made

The Bank of Fayetteville doubled for Ponder Blue's Barbecue on *Evening Shade*. *(Photo courtesy of The Bank of Fayetteville.)*

a bit of a business out of the show's popularity, welcoming tourists, and offering for sale a cookbook with recipes from stars of *Evening Shade* and town residents. The locals hope the proceeds from the cookbook will help fund a new gymnasium/auditorium at the high school.

Actor Burt Reynolds himself visited the town as the featured speaker at the Evening Shade High School commencement, in 1991. Of note is the fact that the school does not, in real life, have a football team.

Shots of country roads and real shade trees in Evening Shade sometimes appeared on *Evening Shade*, but the exterior locations seen most frequently on the show are really in the Arkansas capital city of Little Rock, and in the picturesque northwestern Arkansas resort city of Fayetteville.

The actors on *Evening Shade* did most of their work in a Hollywood studio.

## What to See in Little Rock, Arkansas

- **Wood Newton's house,** 2102 Louisiana Street (at 22nd). This is a private home, known in Little Rock as the Wilson-Mehaffy House, for its earlier occupants. William Wilson, a partner in a grocery business, built the house in 1883. Tom Mehaffy, an attorney and onetime associate justice of the Arkansas Supreme Court, owned the home from 1902 to the mid-1940s. The house, which is located three blocks from the Arkansas Governor's Mansion, is of Italian-influenced design, but the porch, highlighted on the show, is of colonial-revival design and was added in the early 1900s.

## What to See in Fayetteville, Arkansas

- **The Bank of Fayetteville,** 1 South Block (on the square). Producer Harry Thomason dubbed the Bank of Fayetteville the "Barbecue Bank" because on the show, exterior shots of the bank appeared as Ponder Blue's Barbecue. John Lewis, president of the Bank of Fayetteville, is a friend of Thomason's. Fayetteville's historic city square also appeared on the show. For banking information call 501-444-4444.

## What to See in Evening Shade, Arkansas

- **The Information Center Souvenir Shop,** 220 Main Street. Copies of *The Evening Shade Cookbook* and T-shirts that say "Burt Knows Evening Shade," or "Evening Shade Natural State," can be purchased here. Call 501-266-3833. (The souvenir items are also available mail order from the Evening Shade School Foundation, P.O. Box 240, Evening Shade, AR 72532. The cookbook is $7, and the T-shirts are $10, including postage.)

## Fun Facts

- Burt's split with Loni Anderson *(WKRP in Cincinnati)* got many more headlines than the show, with *People* magazine calling it "Hollywood's dirtiest divorce." At first Reynolds blamed the breakup on the long hours he was putting in on *Evening Shade*. But that was before other sordid details were revealed.
- Actor Charlie Dell, who played Nub Oliver, the grown-up

paper boy on the show, chose to get married in the real Evening Shade in Sharp County, Arkansas. He and Jennifer Williams, an actress whose credits include *Cheers*, were married in 1993, in the Methodist Church.

- Veteran actor Ossie Davis, who played Ponder Blue, in real life is married to veteran actress Ruby Dee.
- Marilu Henner, who plays Wood's wife, Ava, earlier was Elaine on *Taxi*.
- Actor Hal Holbrook, who played Evan Evans, publisher of *The Evening Shade Argus*, for several years played Mark Twain in a one-man stage show.

## I'LL FLY AWAY
**(October 1991–October 1993, NBC, then finale and reruns on PBS)**

The setting for this acclaimed southern drama was fictional Bryland, Georgia, in the 1950s, at the beginning of the civil rights movement.

Liberal-minded local DA (and later U.S. attorney) Forrest Bedford (Sam Waterston) is raising his three children, Nathaniel (Jeremy London), Francie (Ashlee Levitch), and John Morgan (John Aaron Bennett), on his own, because his wife is in a mental institution.

He has considerable help in the task from the family's black housekeeper, Lilly Harper (Regina Taylor), who is also raising her own children as a single mother, and, in addition, is involved in seeking racial equality through such activities as voter registration drives and sit-ins.

Produced by Joshua Brand and John Falsey *(St. Elsewhere, Northern Exposure)*, the low-key show was filmed on location in and around Atlanta, with settings generally reflecting a fifties ambiance.

Much of the small city of Madison, Georgia, about an hour's drive east of Atlanta, was featured on the show, including shops and buildings on the city's downtown square.

Tom Pietschner, location manager for the series, said the city has a fifties atmosphere that was "pretty neat" to work in.

Newnan, Georgia, about one hour south of Atlanta, was also

used for several locations, including the house that appeared as the Bedfords' home.

The show also visited the state capitol building and filmed in the real Georgia governor's office in Atlanta.

*I'll Fly Away* won numerous awards and accolades, and had some loyal fans, but it never caught on big with audiences. Some critics expressed surprise when it was renewed by NBC for a second year. Later, PBS decided the show was worthy of another run, and, in addition to showing already aired episodes, ordered up a new two-hour finale.

## What to See in Madison, Georgia

- **Morgan County Courthouse,** Jefferson and Hancock streets (downtown). This 1880s courthouse had been modernized inside, but the show's crew added period detail, replacing fluorescent lighting with old-fashioned school lights and ceiling fans, adding carpeting and wooden window blinds, and restoring the benches. In addition to the interior, seen in courtroom scenes on the show, a lot of shots were done on the street outside.

- **Old Madison Antiques,** 184 South Main Street (downtown). This antique store is in a building that was once a department store, and it appeared as the Pickmore Department Store on the show. It's where lunch counter scenes took place.

- **The Old Carter Pontiac Building,** corner of Hancock and Jefferson. This building appeared on the show as the police station. It really houses business offices.

- **Madison Morgan Cultural Center,** Main Street (about three blocks from downtown). This school, built in 1895, appeared as the school that John Morgan attended on the show.

- **Ye Olde Colonial Restaurant,** East Washington Street (downtown). This southern-style restaurant appeared as a restaurant on the show. Both interior and exterior shots were used. Breakfast, lunch, and dinner are served every day but Sunday. 706-342-2211.

- **Armour's 5&10,** 115 South Main Street. This real-life 5&10 is stocked with period items, which made it a natural to appear on the series.

The Silver Skillet Cafe was one of many real-life locations used on *I'll Fly Away.* *(Photo by Manny Rubio, courtesy of The Silver Skillet Cafe.)*

▶ **Baldwin's Pharmacy,** Main Street. This downtown pharmacy appeared as such on the show.

▶ **The Private House at 557 Academy Street.** This house is one of two that appeared as the home of Christine LeKatzis (Kathryn Harrold), lawyer lover of Forrest Bedford. The other is at 396 East Washington Street.

▶ **Madison Hardware and Supply,** 174 West Washington Avenue. This shop appeared as the bus station on the show.

## What to See in Atlanta, Georgia

▶ **Manuel's Tavern,** 602 North Highland Avenue. Seen on the show as a bar, this neighborhood tavern is a favorite hangout of politicians, including former President Jimmy Carter, as well as the media and average Joe-types too. Bar food is served. Hours of operation are Monday to Saturday, 11 A.M. to 2 A.M. and Sunday, 3 P.M. to midnight. Call 404-525-3447.

▶ **The former Bass High School,** 1080 Euclid Avenue (in the Little Five Points area). This former school building was shown as the school Nathaniel and Francie attended on the show. It's now used as loft space by artists.

▶ **House at 636 Federal Terrace.** This private home appeared as the home of Paul Slocum (Peter Simmons), Nathaniel's wrestler friend, on the show. Interior and exterior shots were used.

▶ **Silver Skillet Cafe,** 200 14th Street. This fifties-style restaurant was used as a restaurant on the show, and has appeared in numerous movies as well. Full-service breakfast here includes country ham and red-eye gravy, and lunch, served weekdays only, is meats and vegetables with corn bread and biscuits. The restaurant is open Monday to Friday, 6 A.M. to 3 P.M.; Saturday, 7 A.M. to 1 P.M.; and Sunday, 8 A.M. to 2 P.M.

▶ **The Point,** 420 Moreland Avenue (in the Five Points area). This venue offers an upstairs bar and a downstairs bar, the latter with live music and dancing. The setting appeared on the show alternately as a black nightclub and a redneck joint. Bar food is served. Upstairs is open 4 P.M. to 4 A.M., daily, and downstairs is open most nights from 9 P.M. to 2 A.M.

## What to See in Newnan, Georgia

▶ **Private House at 38 College Street.** This is the home of the Bedford family on the show, and interior and exterior shots were used. The interior of the house was later re-created on a set in Decatur, Georgia. The street was also shown.

▶ **The Dairy Bar,** 244 Greenville Street. This ice cream shop appeared as The Debbie Dip on the show.

▶ **Mill's Chapel Baptist Church,** 64 Washington Street. The old sanctuary of the church was used on the show.

## What to See in Covington, Georgia

▶ **The private house at 3144 Stone Mountain Street.** This tiny low-rent house appears as Lilly's shack on the show. It was later re-created on the show's set.

### FUN FACTS

- Actor Jeremy London won the role of Nathaniel on the show after his identical twin brother, Jason, turned it down. But it is Jason who played the role in the show's two-hour finale on PBS, because Jeremy was previously committed to another show, *Angel Falls,* on CBS.
- Regina Taylor, who played Lilly, was raised by a single mother in Texas and Oklahoma, and is a graduate of Southern Methodist University in Dallas. While trying to make it as an actress in New York, she once worked as a housekeeper, according to *People* magazine. Her acting credits include the movie *Lean on Me.*
- In the finale, an elderly Lilly (Regina Taylor), still an activist and living in Atlanta, reminisces about the civil rights movement, and her days as a maid. And for the first time she calls Forrest (Sam Waterston) by his first name.
- Sam Waterston went on to play another lawyer, this time of the New York variety, on *Law & Order.*

## IN THE HEAT OF THE NIGHT
**(March 1988–May 1992, NBC and October 1992–May 1994, CBS)**

This cop show was supposed to be set in the small southern town of Sparta, Mississippi, but most of *In the Heat of the Night* was really filmed on location in Covington, Georgia, about thirty-four miles east of Atlanta. (The first few episodes were filmed in Hammond, Louisiana).

Much of the city of Covington appeared on the show, including the courthouse, built in the 1800s, and downtown stores. Most indoor shots were filmed at a local studio.

Fans visiting Covington will find many familiar sites, but the buildings aren't always used the same way as they were on the show. The exterior shot seen as the Magnolia Cafe on the show, for instance, is really an insurance agency. And the Mason Dixon Line Bar is really a beauty shop.

Covington has been successful in drawing large numbers of tourists, with advertising, including billboards, touting its status as a TV location. Maps have been put together to help visitors see the sites used on the show, with copies available at the Hollywood South Souvenir Shop (see below) or Covington City Hall.

Founded in 1822, the city of twelve thousand had been known previously for its historic antebellum-style historic homes and for the local Fox Vineyard Winery, which offers tours and tastings. But *In the Heat of the Night* was big business here, and city officials estimate the show added as much as $4.5 million a year to the city's coffers.

Nearby Conyers, Georgia, also appeared on the show.

*In the Heat of the Night,* which centered on the relationship between a southern white chief of police, Bill Gillespie (Carroll O'Connor), and his northern-trained black chief of detectives, Virgil Tibbs (Howard Rollins, Jr.) was based on the 1967 movie of the same name, starring Sidney Poitier and Rod Steiger, which was based on a novel by John Ball.

## What to See in Covington, Georgia

▶ **Julia A. Porter Library,** 1174 Monticello Street. The library appeared on the show as the Sparta Police Station.

▶ **Office building,** 1108 Conyers Street. This was the home of Virgil and Althea (Howard Rollins and Anne-Marie Johnson) Tibbs on the show. Exteriors only were shown.

▶ **Newton County Courthouse,** 1124 Clark Street. The real-life courthouse, built in the 1800s, appeared as the Sparta Courthouse on the show. It has one main courtroom in the building, and another in an annex out back.

▶ **Covington News,** 1148 Monticello. This real-life newspaper building appeared as the headquarters of the *Sparta Herald* on the show.

▶ **The Covington First Methodist Church,** 1113 Conyers Street. This church appeared as the Sparta Methodist Church on the show.

▶ **The Gary Massey Insurance Agency,** 1159 Monticello. This was the real building seen as the Magnolia Cafe.

The Newton County Courthouse in downtown Covington, Georgia. *(Photo by Marlene Karas.)*

▶ **Private house at 2130 Monticello.** This was where Bill Gillespie (Carroll O'Connor) moved in the 1993 season, when he got married. Only the outside was seen on the show. Earlier, the house seen as Gillespie's was a cabin off Turner Lake Road, about a mile away.

▶ **Building at 2140 Monticello.** Several locations were seen as Bubba's (Alan Autry) apartment, and this is one.

▶ **Sharp Middle School,** 3135 Newton Drive. This real school was Sparta High School on the show.

▶ **Newton General Hospital,** 5126 Hospital Drive Northeast. The back entrance of the real hospital appeared on the show as Sparta General.

▶ **Watties Beauty Shop,** 5101 Floyd Street. This beauty shop's exterior appeared on the show as the Mason Dixon Line Bar.

▶ **Hollywood South Souvenir Shop,** 1160 Monticello Street. This store offers many *In the Heat of the Night* items including auto tags, baseball caps, pins, mugs, china, and photos. Free walking maps are available of sights seen on the show. The shop also offers items by mail order. Call 404-786-2115.

### What to See in Conyers, Georgia

▶ **Michelangelo Restaurant,** 951 Railroad Street (about eight miles west of Covington). On the show, this restaurant appeared as the MacGuffy House Hotel, and has guest rooms, as well as a theater in the back (where Robert Goulet is among those who have performed). But in real life it's a restaurant featuring Italian and Continental cuisine. Interior and exterior shots of the building appear on the show. Call 404-929-0828.

### Fun Facts

• Carroll O'Connor was Archie *(All in the Family)*, a part for which he won four Emmys, before he was Bill, for which he also won an Emmy. O'Connor's real-life son, Hugh O'Connor, played Deputy Lonnie Jamison on *In the Heat of the Night*.

• When Carroll O'Connor had heart surgery in 1989, actor Joe Don Baker came on to play the role of Tom Dugan, a police captain. He earlier starred in the movie *Walking Tall*.

• Quincy Jones helped write the theme song for the show.

• Actor Howard Rollins was earlier in movies including *Ragtime*.

## MAJOR DAD
**(September 1989–September 1993, CBS)**

Gerald McRaney starred in this sitcom as Major John D. "Mac" MacGillis, a military man whose life is changed radically when he meets, falls in love with, and marries Polly Cooper (Shanna Reed), a liberal reporter with three daughters.

On *Major Dad*, Mac works first at fictional camp Singleton in Oceanside, California (where the real Camp Pendleton is located), and, starting in the second season, at fictional Camp Hollister in Farlow, Virginia.

The show was filmed mostly on a set in Hollywood. But shots of the real Marine Corps Base in Quantico, Virginia, were also featured.

In researching their roles, the actors visited the Marine Corps Recruit Depot in San Diego, California, to learn about military procedures. And, according to a spokesman for the Marine Corps, the set designers borrowed items from the Marine Corps to give the base on the show a more realistic feel.

The show incorporated current events affecting the military into its story lines, including the war in the Persian Gulf and military cutbacks.

## What to See in Quantico, Virginia

▶ **Marine Corps Base, Quantico,** Jefferson Davis Highway (off I-95, about thirty-five miles south of Washington, D.C.) The real Lejeune Hall headquarters building here appeared on the show as the base's headquarters. Exterior shots only were used. The town of Quantico, a civilian community, was also featured when Polly (Shanna Reed) ran for mayor, a base spokeswoman said. The base has a museum that is open to visitors, and features displays on the history of ground and aviation forces. It's open Tuesday to Saturday, 10 A.M. to 5 P.M., and Sunday, noon to 5 P.M. Call 703-640-2606.

## Fun Facts

• Actor Gerald McRaney earlier played Rick Simon on *Simon & Simon.* In real life, he is married to Delta Burke

Lejeune Hall at the Marine Corps Base, Quantico. *(Photo courtesy of the United States Marine Corps.)*

*(Designing Women)*. The two visited American troops in Saudi Arabia together in 1991, to show their support for their efforts.

• Vice President Dan Quayle appeared on an episode of *Major Dad* in November 1990, on the 215th anniversary of the Marine Corps.

• Shanna Reed's earlier TV acting credits include *The Colbys*.

## MATLOCK
**(September 1986– , NBC, then ABC)**

Andy Griffith plays Benjamin Matlock, a lovable, grumpy, Harvard-educated lawyer in this crime show, which is particularly popular with older viewers.

The setting for the show is supposed to be the Atlanta area, but *Matlock* was filmed first in L.A., for seven seasons, and then in Wilmington, North Carolina (only the pilot and some exteriors were shot in Atlanta).

Cost-cutting, a change of networks (from NBC to ABC), and actor and executive producer Griffith's desire to be closer to his home (which is a sixty-acre waterfront farm on Roanoke Island, North Carolina, about a fifty-minute plane ride from Wilmington), precipitated the move.

As part of the move from L.A. to Wilmington, the courtroom set for *Matlock* was transported cross-country in seven tractor trailer trucks, and the crew then reassembled the pieces like a giant puzzle at a Wilmington studio, according to location manager Johnny Griffin. The outside of the real U.S. District Court house in Wilmington began appearing on the show in 1993, after production moved south.

Humorously, in shooting scenes in Wilmington, the directors don't necessarily take care to show details that would be Atlanta-ish according to Griffin. For instance, the Cape River in Wilmington could be seen partially in the background in at least one episode, even though downtown Atlanta has no river. The director liked the angle and did not want to change the setting to eliminate the water view, Griffin said.

Some filming for the show also took place in Southport, an historic fishing village about twenty-seven miles south of Wilmington, where the Cape Fear River meets the Atlantic Ocean.

## WHAT TO SEE IN WILMINGTON, NORTH CAROLINA

▶ **The U.S. Customs House, Alton S. Lennon Federal Building,** 2 Princess Street (at North Water Street, between Market and Princess). The real home of the federal court in Wilmington, this block-size building of neoclassical design appears as the courthouse on *Matlock,* too. The building dates back to 1919, and is on the National Register of Historic Places. It also houses the local offices of federal government agencies.

## WHAT TO SEE IN SOUTHPORT, NORTH CAROLINA

▶ **The private house at 106 Lord Street** (off of Moore Street). This white clapboard, Southern-style private house is seen as Matlock's on the show. The front porch was re-created on the *Matlock* stage.

Visitor information on historic Southport can be obtained from the local Chamber of Commerce. Call 910-457-6964.

## FUN FACTS

• On the show, the character of Benjamin Matlock (Andy Griffith) will eat almost anything, including peanut butter and mayo sandwiches with pickles and bananas.

• In a 1993 survey, the *National Law Journal* found that a third of Americans believe that lawyers are less honest than the rest of the population. Asked to name an attorney they most admire, more than half didn't answer, but the other half included Matlock and Perry Mason on the list.

• In 1988, the show's producers let the audience decide the ending on one episode, having viewers vote by calling a 900 number. Three separate endings were filmed.

• Linda Purl, who plays Matlock's lawyer daughter, Charlene, earlier played Fonzie's girlfriend on *Happy Days.* In 1993, she married a real British peer, Lucius Alexandar Plantagenet Cary, the Master of Falkland.

• Actor Don Knotts (Barney Fife on *The Andy Griffith Show)* was reunited with his former partner, Griffith, on *Matlock,* playing Matlock's neighbor, Les Calhoun.

## THE WALTONS
**(September 1972–August 1981, CBS)**

"Good night, John-Boy . . ."

Series creator Earl Hamner, Jr., was one of eight children brought up in a rural Virginia family during the Great Depression, and his real family's life is the basis for this heartfelt family show. Hamner based the character of John-Boy on himself, and the role made actor Richard Thomas a star.

The family was first introduced in a TV movie, *The Homecoming* (starring different adults but the same child actors as the show), which ran as a Christmas special in 1971. So positive was the public response to the quiet country clan, that the subsequent series was developed.

*The Waltons* was shot in Hollywood. The house seen on the show, however, was a re-creation of the real house in the Blue Ridge Mountains town of Schuyler, Virginia, where Hamner and his brothers and sisters were raised.

So engaging was the Walton family that many fans went searching for the show's fictional Walton's Mountain.

Hamner wrote in *TV Guide* in 1993 that tourists who managed to find his real hometown of Schuyler during the show's run did not go away disappointed. The demand proved a bit much for his mom, however, as she felt it her duty to show southern hospitality by offering the visitors tea. Eventually a fence was put up around the property.

Hamner's parents have now passed away, and as of 1993, the house was occupied by James Hamner, one of Earl's brothers, a computer analyst, and the inspiration for Jim-Bob (David Harper) on the show.

In response to continued tourist interest, Walton's Mountain Museum opened in Schuyler, in 1992, across the street from Hamner's boyhood home.

The tiny central Virginia town of four hundred is located about thirty miles from Charlottesville, home of Thomas Jefferson's Monticello.

## What to See in Schuyler, Virginia

▶ **Walton's Mountain Museum,** Route 617. Recognizing that writer Earl Hamner was a great local asset, local residents developed this museum in an old school building. Money raised through the endeavor is used to fund a community center. The museum has five rooms, with the living room and kitchen designed to simulate those seen on the show. John-Boy's room is decorated with artifacts and memorabilia from Earl Hamner's life, and is designed to show visitors the environment in which he created. There is also a re-creation of Ike Godsey's general store, but this one sells local crafts and souvenirs. And there is an audio/visual room where visitors can hear recordings of cast members reminiscing about the show. The museum is open from the first Saturday in March through the last Sunday in November (with the exception of Easter, Thanksgiving, and the second Saturday in October), 10 A.M. to 4 P.M. Admission is charged. Call 804-831-2000.

▶ **The real Hamner home,** Route 617. The actual home is a private residence, but tour guides from the museum can point it out for exterior viewing only.

The home of the Hamner family was the model for the house seen on *The Waltons. (Photo courtesy of The Waltons Mountain Museum.)*

### Fun Facts

• Series creator Earl Hamner had earlier written scripts for *The Twilight Zone.* He later developed *Falcon Crest.*

• Even before *The Homecoming,* Earl Hamner wrote a script based on his boyhood for the movies. The result was *Spencer's Mountain,* which starred Henry Fonda, Maureen O'Hara, and James MacArthur, and was set in Wyoming.

• John Ritter *(Three's Company, Hearts Afire)* appeared from 1973 to 1975 on the show as Reverend Fordwick, who married the schoolteacher, Miss Hunter (Mariclare Costello).

• Actress Ellen Corby (Grandma) suffered a stroke in the fall of 1976 and missed a year and a half of the series, returning in 1978. Will Geer (Grandpa) died before the 1978 season began. The actor was not replaced, and the death of Grandpa was mourned on the show.

## FYI: ALSO IN THE SOUTH

The George Lucas-produced show, *The Young Indiana Jones Chronicles* (March 1992 to July 1993, ABC) was filmed mostly in North Carolina, with the show's headquarters in Wilmington.

Location shoots were done in several places including Raleigh, Durham, and Chapel Hill, but the locations were always supposed to be someplace else.

The North Carolina governor's mansion, for instance, doubled for the famous Algonquin Hotel in New York (home of the legendary Round Table).

### What to See in Raleigh, North Carolina

▶ **The North Carolina Executive Mansion,** 200 North Blount Street. The main entrance hall area, which has Corinthian columns, grand staircase, ballroom, library, and ladies parlor were all shown on the show. The rooms were redecorated by the set designers during the filming. The mansion, which dates from the 1880s, is open to visitors for tours on a limited basis, generally on Tuesday, Wednesday, or Thursday. Reservations are recommended. Call 919-733-3456.

# FLORIDA

*Miami Vice* is, of course, Miami's main claim to TV fame. But as familiar as the sunny scenery, blue waters, and pastel-colored buildings became to viewers of the cop show, things almost didn't happen that way.

Producer Michael Mann actually had first considered basing the show in New York. Would Sonny have had to wear socks?

*Miami Vice* was shot on location, in Miami, and visitors can still find the hip haunts of Sonny and Tubbs in and around the city.

The comedy/drama *Key West* was supposed to do what *Miami Vice* did for Miami, put the city higher on the resort map, but Fisher Stevens's show was canceled (even though he got a lot of press for dating Michelle Pfeiffer), after eleven weeks.

Still, the show showed some nice location footage including frequent shots of Smathers Beach. And while one local said there really aren't as many off-beat characters in the area as on the show, she sounded a bit off-beat herself.

*Flipper,* that most famous of dolphin shows, was a Florida production, with water scenes done at the real Miami Seaquarium. Fans can reminisce about the show today while watching dolphins in the attraction's live *Flipper* show.

*I Dream of Jeannie* shot exteriors at the real Patrick Air Force Base in Cocoa Beach, including the building with the rockets out front, seen as the workplace of astronauts Tony and Roger.

*The Golden Girls* were single ladies of mature age enjoying life in Florida, on the popular sitcom. But the show was really shot in L.A., as was its follow-up show, *Golden Palace*.

## FLIPPER
**(September 1964–September 1968, NBC)**

On this family show, Porter "Po" Ricks (Brian Kelly) is the chief ranger at Coral Key Park in Florida. And his job is to protect sea creatures, as well as humans, from dangers in the waters at the facility.

Ricks gets considerable help in the task from Flipper, a friendly dolphin who is the pet of his children, Sandy (Luke Halpin) and Bud (Tommy Norden). Flipper lives in the lagoon outside the Rickses' cottage.

The real-life setting seen as Coral Key Park in one hundred episodes of *Flipper* (as well as in the pilot and two movies) is the lagoon at the Miami Seaquarium. The Bahamas was also used for location shoots.

Today, visitors to the Miami Seaquarium can see a re-creation of ranger Ricks's house as well as the dock area. And in the attraction's lagoon, the tradition of *Flipper* is preserved, with a show starring dolphins trained in similar behaviors to those seen on the show.

Jose Hoyo, the Miami Seaquarium's general manager, said the original ranger house was a Hollywood stage prop so it was not built to code and eventually had to be demolished. The new house, built at the original location, is used by the real-life dolphin trainers as a locker room and food preparation area.

The marine mammals seen on *Flipper* are long gone, but have been replaced in the Miami Seaquarium's show by other dolphins, all born at the facility and trained to display similar behaviors as those seen on the show. All are Atlantic bottle-nosed dolphins, found off the Florida coast.

The behaviors take many months, or even years, to learn, but are all based on things the dolphins would do in the wild; a sort of dolphin aerobics display.

"The animals actually like to do this display. To them it is exercise," Hoyo said.

### What to See in Key Biscayne, Florida

- **Miami Seaquarium** and **The Flipper Lagoon,** 4400 Rickenbacker Causeway. There's a Flipper show as well as other shows starring Lolita the Killer Whale and other sea creatures. Blown-up photos from *Flipper* are displayed. Admission is charged and souvenirs are for sale. Open daily at 9:30 A.M. Call 305-361-5705.

## FUN FACTS

- The part of Flipper was credited to a dolphin named Suzy, but according to officials at the Miami Seaquarium, several dolphins were actually shown in the part. Flipper was supposed to be a boy dolphin.
- Tom Norden, who played Bud, went on to play Gary Walton on *Search for Tomorrow,* but later gave up acting to become a businessman. According to *People* magazine, he runs a headhunting firm in New York.
- Luke Halpin, who played Sandy, continues to work on marine film projects, but behind the scenes as a crew member

The *Flipper* show at the Miami Seaquarium. *(Photo by Lourdes Segrera, courtesy of the Miami Seaquarium.)*

rather than in front of the cameras. According to *People* magazine, he lives in Florida.

• The show was created by Ivan Tors, who also produced the outdoors shows *Sea Hunt*, *Gentle Ben*, *Daktari*, and *The Aquanauts*.

## I DREAM OF JEANNIE
**(September 1965–September 1970, NBC)**

Astronaut Tony Nelson (Larry Hagman) crashes a NASA rocket on a deserted island in the South Pacific and finds an antique bottle. He opens it and out pops Jeannie (Barbara Eden), a blonde in a harem costume, with magical powers.

The two return to Cocoa Beach, Florida, thanks to Jeannie's powers, where they at first live platonically (for four years), and eventually fall in love.

Reality merged with fantasy in 1969, when then-Florida Governor Claude Kirk and the local Chamber of Commerce in Cocoa Beach hosted a real four-day celebration in honor of the wedding of Jeannie and Tony Nelson. And some of the events are seen on the show.

The cast toured the nearby NASA and Air Force facilities and visited Cocoa Beach City Hall to take out a mock wedding license application (on the show, Jeannie's genie blood type causes complications in getting the necessary health certificates for the wedding).

*I Dream of Jeannie* was mostly shot in Hollywood, but actual footage was shown of rocket launches at Cape Kennedy in Florida, as well as at the Manned Space Center in Houston. Downtown Cocoa Beach was seen on the show as well, as was Patrick Air Force Base.

The show's producer, Sidney Sheldon (who became a popular novelist) spent time in Cocoa Beach (his uncle lived there), and references to local businesses and attractions are made on the show.

The Cocoa Beach area has so much real-life space activity, as home to the space shuttle program, it is known as Florida's Space Coast. Visitors might not see Tony, but they'll find plenty of astronaut-related attractions to visit.

## What to See in Cocoa Beach, Florida

▶ **Patrick Air Force Base,** on A1A (south of downtown Cocoa Beach). The building that appeared on *I Dream of Jeannie* as the place where Tony and the other astronauts worked is really the Air Force Technical Applications Center headquarters here, on the south end of the base. There used to be nine rockets or missiles in front, but six have been removed due to rusting. While there was some NASA activity at the base at the time of the show, Air Force business with NASA now takes place at the Cape Canaveral Air Force Station to the north.

▶ **Bernard's Surf Restaurant,** 2 South Atlantic Avenue. Rusty Fischer, the owner of this seafood restaurant, says the name of the restaurant was mentioned on the show quite a bit. And he hosted the cast and crew for dinner when they were in town.

Rockets and missiles at Patrick Air Force Base. *(U.S. Air Force photo.)*

Real-life astronauts also have been known to frequent the award-winning restaurant. It's open daily, for lunch and dinner. Call 407-783-2401.

## What to See at the Kennedy Space Center, Florida

▶ **Spaceport USA,** off State Road 405 (just north of Cocoa Beach, and about fifty-five miles from Orlando). The center offers guided tours including real Space Shuttle launch sites, exhibits on space exploration, and demonstrations such as a simulated Apollo 11 moon launch countdown. A large gift shop offers 1,200 space-related souvenirs. An IMAX theater offers two giant-screen space movies. The Astronaut Memorial is dedicated to the astronauts who have given their lives for space exploration. There is also a Rocket Garden that offers a collection of eight rockets from past eras. The center is surrounded by a seventy-acre wildlife preserve with alligators, armadillos, and other native creatures. Tours begin daily at 9:45 A.M., and end two hours before dark. Call 407-452-2121.

▶ **Air Force Space Museum,** Space Launch Complex 26. At the site where the first satellite, *Explorer 1*, was launched, the museum has on display 709 missiles and space-launch vehicles. It is open on Sunday, except when launch operations are in progress. Call 407-494-NEWS.

▶ **Astronaut Hall of Fame,** State Rd. 405 (just west of the Kennedy Space Center). Alan Shepard, Wally Schirra, Scott Carpenter, and John Glenn are among the real-life astronauts lending their faces and voices to the exhibits in this museum. Visitors can participate in a multi-media space shuttle flight; seat belts are required. The museum is also home to the U.S. Space Camp, open to kids in grades 4 to 7. The museum is open daily except Christmas Day, 9 A.M. to 5 P.M. Call 407-269-6100.

For information on area attractions call the Florida Space Coast Office of Tourism, 800-USA-1969.

## Fun Facts

• The Blue Djin (Michael Ansara), the most powerful and most feared of genies, turned Jeannie (Barbara Eden) into a genie more than two thousand years ago, when she spurned his marriage proposal.

• The show's creator, Sidney Sheldon, earlier created *The Patty Duke Show,* and went on to create *Hart to Hart.* Sheldon hit it big as a novelist too, and some of his books, including *If Tomorrow Comes* and *Master of the Game,* were made into TV miniseries.

• Jeannie was originally envisioned as a brunette.

• Barbara Eden was pregnant with her son, Matthew, when the first thirteen episodes of the show were shot in 1965. Veils helped her cover that fact.

• Larry Hagman said he took the part of Tony "for money." He later said the same thing about his role as J.R. Ewing *(Dallas).*

• Jeannie's navel was prudently not shown on the original series. But it did appear in a 1985 reunion movie.

• Larry Hagman did not appear in the reunion movie because he was better known by then as J.R. of *Dallas.* Wayne Rogers *(M*A*S*H)* plays Tony instead. Eden also starred with Rogers on stage in *Same Time, Next Year.*

• Barbara Eden also played Jeannie's brunette sister, Jeannie II, when she appeared on the show.

• *I Dream of Jeannie* started in black and white, and switched to color in 1966.

• A cartoon version of *I Dream of Jeannie,* called *Jeannie,* was introduced in 1973.

## MIAMI VICE
**(September 1984–July 1989, NBC)**

They were hip. They were cool. They were MTV-cops. And hey, they weren't bad to look at either.

Sonny Crockett (Don Johnson) and Ricardo Tubbs (Philip Michael Thomas) introduced the world to the no-socks look for men and to sunny, beautiful Miami as well.

When the show was first scheduled to air, some residents of Miami feared it would unfairly depict the city only as a home to drug dealers and the underworld. But their fears were soon set aside. *Miami Vice* did for Miami what *The Love Boat* did for the cruise industry, drawing a whole new tourist audience.

Despite the crime plots, the cop show actually showed Miami as a modern city with beautiful young people, great weather, and exciting night life. And as a result of the show, the city's reputation of being mainly a resort to the elderly ("God's waiting room") dissipated. Actors Don Johnson and Philip Michael Thomas became local heroes, and body guards had to be hired to keep away the crowds of fans.

The show was filmed on location in Miami, and many sites in the city will look familiar to fans. One site that can't be seen anymore is Miamarina, where Crockett's boat was docked on the show. It was closed after Hurricane Andrew.

## What to See in Miami Beach, Florida

- **Ocean Drive in Miami Beach,** from about 5th Street to 12th Street (including the historic Art Deco District). This South Beach area is a hip part of the city referred to by locals as SoBe (a takeoff on New York's SoHo). SoBe frequently appeared on *Miami Vice,* and was also featured on the short-lived *Moon Over Miami.* Also seen on *Miami Vice* was the nearby Lincoln Road Mall, a pedestrian area of upscale shops, theaters, and art galleries, and Espanola Way, with its Spanish-style architecture, shops, and nightclubs.

- **The Alexander All-Suite Luxury Hotel,** 5225 Collins Avenue. This pink hotel, smack in the middle of Miami Beach, was the headquarters for production of *Miami Vice,* and also appeared in several episodes. Don Johnson lived in one of the hotel's suites, and the cast and crew frequented the hotel's upscale French restaurant, Dominique's, in the process putting the establishment on the map. The restaurant still offers a Don Johnson dessert, a pistachio souffle. Parking spaces in the garage still have reserved signs with *Miami Vice* on them. The hotel has 160 suites ranging from one bedroom to a large penthouse. Rates for one-bedroom suites begin at $225 in the summer and $310 in the winter. For reservations call 305-865-6500.

- **Star Island.** This posh residential area is home to several private houses that appeared on the show, including one made of chrome and glass. The island is guarded by a private gatekeeper, so access is limited.

Downtown Miami, the setting for action on *Miami Vice*. *(Photo courtesy of the Miami Convention & Visitors Bureau.)*

## What to See in Miami, Florida

▶ **The Miami Shipyard Building,** Southwest 2nd Avenue (at Southwest 8th Street). This limestone Art Deco building appears as the police headquarters on the show. It was once used to house a refurbishing operation for boats, but is now home to some charter boat companies.

▶ **Biscayne Boulevard** (from the southern end to around 13th Street). This is where chase scenes for the show were often filmed.

▶ **The Atlantis,** 2025 Brickell Avenue. This architecturally unusual building is seen in the opening sequences of the show. It has a palm tree, Jacuzzi, and red spiral staircase visible several stories up.

▶ **Miami Jai-Alai,** 3500 Northwest 37th Avenue, near Miami International Airport. This is the jai alai fronton seen in the show's

The Atlantis building in Miami. *(Photo by Norman McGrath, courtesy of Arquitectonica.)*

intro, and it's also the nation's oldest fronton. Open 7 P.M. to midnight, except Tuesday and Sunday. Matinees are offered on Monday, Wednesday, and Saturday. Call 305-633-6400.

- **Vizcaya Museum,** 3251 South Miami Avenue (in the Coconut Grove section). The exterior and formal gardens of this renaissance-style palace were seen on the show. Built in 1916 as a winter retreat for industrialist James Deering, the house contains Deering's personal collection of European decorative arts. Pope John Paul II and President Reagan met here in 1987, during the pontiff's U.S. tour. Open daily, except on Christmas Day. Call 305-579-2708.

- **Parrot Jungle and Gardens,** 11000 Southwest 57th Avenue. The flock of flamingos seen in the show's intro was filmed here. Performances by trained birds, flamingo feeding demonstrations, a bird sanctuary, a petting zoo, a baby ape exhibit, and a playground are among the offerings at this attraction. There's also a gift shop. Open daily 9:30 A.M. to 6 P.M. Admission is charged. Call 305-666-7834.

The famous flamingos at Parrot Jungle and Gardens. *(Photo courtesy of Parrot Jungle and Gardens.)*

## Fun Facts

- The MTV pace of the show was highlighted by guest stars who included rockers Glenn Frey, Ted Nugent, and Phil Collins. Singer Sheena Easton played rock star Caitlin Davies, who marries Crockett but is later killed off. Also appearing on the show were music greats James Brown and Little Richard.
- The music on *Miami Vice* was mostly created by Jan Hammer, with contemporary artists featured. It did not come cheaply, with up to $50,000 spent per episode for the background score. The show's theme song, by Hammer, hit number one on the pop charts, and the show also spawned two record albums.
- Melanie Griffith, Don Johnson's then ex-wife, also guest-

starred on the show. The two rekindled their relationship and were married for a second time. They later made movies together including *Paradise* and *Born Yesterday.* A recovering alcoholic, Johnson had a relapse in 1994, and he and Griffith parted again.

• Sonny Crockett (Don Johnson), whose first name was really James, drove a Ferrari and had a pet alligator named Elvis. He lived on a boat, and frequently dressed in sport coats with T-shirts underneath.

• Ricardo Tubbs (Philip Michael Thomas) was an ex-New York cop, who came to Miami, originally, to track down the drug dealer who killed his brother. He adapted well to Miami ways, driving a vintage Cadillac and wearing chic Armani suits.

• In the pilot for the series, Sonny's partner, who was killed, was played by Jimmy Smits *(L.A. Law, NYPD Blue).*

• When the popular show was canceled, it was due in part to the cost of production. *Miami Vice* cost as much as $1.2 million per episode to produce.

# THE MIDWEST

While America's Heartland has been well-represented in TV-land, most of the shows that have featured Midwest settings in their plots really have been shot in Hollywood.

That's not to say there aren't some wonderful real Midwest treats for TV fans to view.

That fabulously funky house that Mary Richards called home on *The Mary Tyler Moore Show* really exists in Minneapolis, as does the apartment building where Mary later moved (located near the University of Minnesota campus). And visitors to the city can throw their hat in the air like Mary at the very real Nicollet Mall area downtown.

In more recent years, Minneapolis is where Brenda first headed to college on *Beverly Hills 90210* (although she later transferred to a California campus because people in Minnesota asked her too many questions about life in Beverly Hills). The hit teen show filmed briefly at the real University of Minnesota campus (showing the Northrop Mall, among other sites).

In Minneapolis you can buy a Minnesota State Screaming Eagles T-shirt, but there is no real college with that name. It's the fictional college on *Coach*.

The sitcom really does, however, feature shots of another college campus, that of the University of Iowa, in Iowa City.

Also in the Hawkeye State, believe it or not, is the only permanently displayed replica of the starship *Enterprise* from *Star Trek*. It's in Riverside, Iowa, which proclaimed itself the future hometown of Captain Kirk after local officials discovered that series creator Gene Roddenberry had written that the character of Kirk was born in a small town in Iowa.

Iowa was also the home state of Donna Stone on *The Donna Reed Show*, and actress Donna Reed hailed from Denison, Iowa, in real life too.

Milwaukee's place in TV history belongs to *Happy Days*, and its spin-off *Laverne & Shirley*. Fans of the nostalgic *Happy Days* can still eat at the kind of old-fashioned drive-ins that inspired the show. And *Laverne & Shirley* fans can imagine the wacky pair at work by taking a brewery tour, or visiting a neighborhood that looks an awful lot like where the duo lived on the show.

Also in the Badger State is Rome, the quiet setting for *Pickett Fences*. The real village (which is too small to be a town) is seen in snippets in the intro to the small-town drama.

*One Day at a Time*, the popular single parent sitcom, was based in Indianapolis, Indiana, but there is no record of the show ever filming in the Hoosier State. Similarly, *Family Ties* did not actually film in Columbus, Ohio, although the show did use props from Ohio's capital city, and made references to real places in Columbus too.

*WKRP in Cincinnati* showed some shots in its intro of downtown Cincinnati, but only one episode was done on location in Ohio, and that was at Paramount's Kings Island in Mason. The amusement park was also the setting for episodes of *The Brady Bunch* and *The Partridge Family.*

*Mary Hartman, Mary Hartman*, the soap opera spoof, took place in Fernwood, Ohio, but the town is fictional.

*The Beverly Hillbillies* really went on location to Missouri's Ozark mountains for several episodes, when the Clampetts left California to visit their hometown. Filming was done at the historic buildings of Silver Dollar City, outside of Branson, Missouri. And at the College of the Ozarks, also outside of Branson, TV fans can view the real rustic car the Clampetts drove on *The Beverly Hillbillies*, which is on permanent display at the Ralph Foster Museum.

*The John Larroquette Show* is shot in L.A., not St. Louis, but *Lucas Tanner*, the teacher show starring David Hartman *(Good Morning, America)*, really did film some exteriors in the St. Louis suburb of Webster Groves, Missouri.

## COACH
**(February 1989– , ABC)**

Producer Barry Kemp has denied there are any similarities between fictional Minnesota State and his alma mater, the University of Iowa, but others say it seems more than coincidence that the coach on the show is Hayden Fox (Craig T. Nelson), and the real-life coach of the U of I is also named Hayden, as in Hayden Fry.

Fry has actually been mentioned on the show a few times. In one episode, for instance, Fox attends a coaches' conference and finds the hotel doesn't have a room for him because they have mixed him up with Coach Fry. Other splashes of references to the U of I have also appeared on the show, say insiders at the school. And you don't have to look too deep. The fictional team (the Screaming Eagles) is, after all, named after a bird of prey, as are the real Iowa Hawkeyes.

Besides his collegiate connection, the main reason Kemp chose the U of I for exterior location shots appears to be the Midwestern ambiance of the 26,000-student campus. Campus sites most often seen on the show are exteriors of Hillcrest Dormitory, a coed dorm, which appears as Coach Hayden Fox's office location, and the University of Iowa Field House, which is seen as a setting for classes. A busy intersection outside Hillcrest Dormitory is also featured in scenes showing students heading to classes.

Among other campus features that have appeared on the show is the scoreboard at the football field. The Iowa Hawkeyes symbol on the scoreboard was digitally changed in one episode, to show a pineapple, when the fictional Screaming Eagles team appeared in a bowl game in Hawaii.

The show is actually filmed on a set in L.A., and the main actors have not done scenes at the U of I campus.

Kemp, however, has been back to the school since graduating in 1971 with a bachelor of arts degree in theater arts. He and his wife, Margaret Gomez, a 1974 graduate, have been ardent supporters of the theater arts program and established a scholarship fund to assist promising playwrights. They also help underwrite the cost of the annual spring University of Iowa Playwright's Festival. In 1991 Kemp received the university's Distinguished Alumni Achievement Award.

For the record, the Iowa Hawkeyes football team has a much better record historically than the Screaming Eagles. Although 1993 wasn't a great year, most other years under Coach Fry have been impressive, and the team has appeared in at least eleven bowl games under Fry's leadership.

## What to See in Iowa City, Iowa

- **Hillcrest Residence Hall,** Grand Avenue, near the center of the campus, west of the Iowa River. The dorm overlooks the Iowa River and is home to 880 students. It is also home to the Foreign Language House, which offers undergraduate and graduate students cultural and language learning opportunities in the form of such events as festivals and film and video showings.

- **The Field House,** Grand Avenue, near the center of the campus. Once used for Iowa Hawkeye basketball games, which are now played at the large Carver-Hawkeye Arena, the Field House is now the main campus recreation facility. The building is open from 8 A.M. to 10:30 P.M. daily when school is in session. Hours vary during the summer and holidays. For more information call 319-335-9293.

The University of Iowa campus, as seen on *Coach*. *(Photo by John Van Allen.)*

▶ **Kinnick Stadium,** University of Iowa campus. The real stadium is named after Nile Kinnick, the only Hawkeye to win a Heisman trophy (1939). It's where the Hawkeyes football team plays.

▶ While there is no Touchdown Club, as on the show, there are a few bars and restaurants with a sport ambiance in Iowa City. Insiders say the best sports bar is the *Field House,* 111 E. College Street, 319-338-6177. But if you want to eat while you drink, a better bet is *The Airliner,* 22 South Clinton Street, 319-337-5314, a restaurant that is particularly popular with fraternities and is partially owned by Brad Lohaus, a former Hawkeye basketball player, who later played for the NBA.

For general campus information and event listings at the University of Iowa, call the *Iowa Memorial Union,* the campus student union, 319-335-3055.

For other attractions in the area, contact the Iowa City and Corallville Convention and Visitors Bureau, 319-337-6592, or 800-283-6592.

## Fun Facts

• Actor Craig T. Nelson attended the University of Arizona. An actor for many years, at one point he sought an alternative life, giving up Hollywood for Mount Shasta, North Carolina, a rural area where he worked as a janitor, a teacher, a land surveyor, and a carpenter, before returning to Hollywood and *Coach* success.

• Jerry Van Dyke (Luther), brother of Dick, is a stand-up comedian who got his television start on *The Dick Van Dyke Show.*

• Bill Fagerbakke (Dauber) attended the University of Idaho on a football scholarship, but a knee injury ended his career. He named his family dog Shelley, for his co-star.

• Shelley Fabares (Christine) is the only actress to have costarred three times with Elvis Presley. She appeared in *Clambake, Spinout,* and *Girl Happy.*

• The niece of Nanette Fabray, Shelley Fabares was married in 1984 to Mike Farrell (Captain B. J. Hunnicutt on *M*A*S*H*).

## THE DONNA REED SHOW
**(September 1958–September 1966, ABC)**

Donna Reed was born Donna Belle Mullenger in a two-story farmhouse seven miles from Denison, Iowa.

The house is no longer there, but to the residents of Denison, the actress, who played all-American housewife Donna Stone on *The Donna Reed Show,* and also won an Oscar for her performance in *From Here to Eternity,* remains a hometown hero. *The Donna Reed Show* was filmed in Hollywood and took place in the fictional small town of Hilldale. But on the show, Donna Stone was, like the star herself, from Denison.

Reed, who died in 1986, is remembered today by the Donna Reed Foundation for the Performing Arts, which sponsors the annual Donna Reed Festival in Denison. The week-long festival, held in June, attracts Hollywood celebrities, which have included cast members from *The Donna Reed Show* and features events and workshops on directing, acting, and writing. The foundation also offers college scholarships to students of the performing arts.

Reed is also remembered in town with Donna Reed Road, located near the golf course. The actress's 1955 Oscar for *From Here to Eternity* is housed in the McHenry House Museum. In addition to acting, Reed was the uncredited producer of *The Donna Reed Show* and directed many episodes herself.

### What to See in Denison, Iowa

▶ **McHenry House Museum,** 1428 First Avenue North. This historic home is run by the Crawford County Historical Society, and houses Donna Reed's Oscar.

▶ **The Donna Reed Foundation.** This is the place for information on the Donna Reed Festival and related events. Call 800-336-4692 or 712-263-3334.

### Fun Facts

• Donna Reed was also familiar to TV fans as the actress who replaced Barbara Bel Geddes as Miss Ellie on *Dallas.* Bel

Geddes left *Dallas* in 1984, after undergoing open heart surgery, but returned the following year. Reed sued the show's producers, Lorimar Productions, because she had a two-year contract to play the role. She was paid more than $1 million in an out-of-court settlement.

- Bob Crane played neighbor Dave Kelsey on *The Donna Reed Show.* He left the show to star in *Hogan's Heroes.*
- Shelley Fabares *(Coach)* played Donna's daughter, Mary Stone. In one episode Mary sang "Johnny Angel," and the song was made into a hit record by Fabares.
- Paul Petersen, who played Donna's son, Jeff Stone, had been an original Mouseketeer on *The Mickey Mouse Club.* When Donna and Alex Stone (Carl Betz) take in an orphan in the 1963 season, the part of the orphan, Trisha, is played by Petersen's real sister, Patty.
- Donna Reed, who died of cancer in 1986, was a cofounder and cochairman of Another Mother for Peace, a national organization opposed to the politics that brought the U.S. into the Vietnam War. She also actively opposed nuclear energy and nuclear weapons.

## FAMILY TIES
**(September 1982–September 1989, NBC)**

The interplay between parents and their kids, and 1980s values were the subjects of this popular sitcom. The show made a major star of Michael J. Fox, who played Alex Keaton, the conservative Republican son of liberal parents living in Columbus, Ohio. Alex venerated Richard Nixon and William F. Buckley, Jr., and money was his first true love. He eventually left home for Wall Street.

Alex's parents, Elyse (Meredith Baxter Birney) and Steven (Michael Gross), were former flower children who went to Woodstock (Elyse gave birth to Alex in a commune). Steven works for the local public TV station, and Elyse is an architect.

Their other kids are Mallory (Justine Bateman), who loves going shopping; Jennifer (Tina Yothers), a bright tomboy; and young Andrew (Brian Bonsall), who was born in 1985 and is a clone of his big brother, Alex.

On the set of *Family Ties,* Meredith Baxter-Birney (right) wears an apron from Max & Erma's in Columbus, Ohio. *(Photo courtesy of Max & Erma's Restaurants.)*

While the show never filmed in Columbus, the capital city of Ohio, the producers took care to establish local color by naming real Columbus locations, even requesting props from local establishments. Cast members on the show, for instance, sported the popular "Surf Ohio" T-shirts, favored by some Columbus locals.

## What to See in Columbus, Ohio

▶ **Max & Erma's Restaurant, Bar & Gathering Place,** 739 South Third Street (in the German Village neighborhood). Elyse Keaton (Meredith Baxter Birney) sometimes wore an apron from this restaurant on the show. The Columbus-based chain offers gourmet hamburgers, onion rings, homemade pasta, salads, and an ice cream sundae bar. This location is one of five in the Columbus area. Open Monday through Saturday, 11 A.M. to 1 A.M., and Sunday 11 A.M. to midnight. Call 614-444-0917.

▶ **Lazarus Department Store,** Town and High Streets. This department store chain was one of the places where Mallory liked to shop on the show. The main store is the anchor store of the Columbus City Center mall. The store is open Monday through Saturday, 10 A.M. to 9 P.M., and Sunday, noon to 6 P.M.

▶ **WOSU Public Television and Radio Station,** 2400 Olentangy River Road (on the Ohio State University campus). This is the real public television station in Columbus. On the show, Steven Keaton worked for fictional WKS. Tours are offered on an appointment-only basis. Call 614-292-9678.

## Fun Facts

• The series was created by Gary David Goldberg, who later based his series *Brooklyn Bridge* on his own childhood.

• Tracy Pollan, who played Ellen Reed, Alex's first college girlfriend at Leland College, in real life married Michael J. Fox.

• Before he became a big movie actor, Tom Hanks appeared on the show as Elyse's brother, Ned O'Donnell.

• When Elyse was pregnant on the show in 1984, actress Birney was also really pregnant with twins.

• The role of Andrew was originally played by babies, but after the summer break the character appeared as a preschooler, played by Brian Bonsall.

• Meredith Baxter-Birney earlier was Nancy on the drama *Family.* She appeared with her then husband, David Birney, on the short-lived comedy *Bridget Loves Bernie* (she as Bridget and he as Bernie). In real life the couple divorced in 1989.

• In a bit of experimental television, Michael J. Fox did a poignant one-man episode of *Family Ties* in which character Alex dealt with the death of a friend in a car accident.

• Canadian-born Michael J. Fox went on to become a movie star, including his role in the phenomenally successful *Back to the Future* movies.

• According to *TV Guide,* Actor Michael Gross is a railroad buff, who bought a hundred miles of real rail track in New Mexico that was in danger of destruction.

• Fans of Alex included President Ronald Reagan, who was once quoted as saying *Family Ties* was his favorite show.

## HAPPY DAYS
**(January 1974–July 1984, ABC)**

"Love and the Happy Day," a skit on *Love, American Style*, about an all-American 1950s family getting their first TV set, inspired George Lucas's hit movie *American Graffiti*. The popularity of the nostalgic 1950s movie in turn inspired *Happy Days*, which became a number one TV show.

Ron Howard starred in the sitcom as straight arrow Richie Cunningham, who with his best friend, Potsie Weber (Anson Williams), attended Jefferson High in Milwaukee. The two hung out at Arnold's Drive-In, as did Fonzie (Henry Winkler), a leather-clad but sensitive biker who became the runaway superstar of the show, even though the character was originally planned as a minor player.

While the setting for the sitcom was Milwaukee, the show never actually did any filming in the city, according to local sources. The house seen as the home of the Cunningham family is really a private residence in Hollywood, and the drive-in seen as Arnold's was also in L.A. But the writers on the show included native Milwaukeeans, and there were plenty of references made on *Happy Days* to the city. Members of the cast visited Milwaukee in the early 1980s, to a hero's welcome. And actor Tom Bosley, who played Howard Cunningham, accepted a key to the city.

Several real-life Milwaukee drive-ins have over the years claimed to be the inspiration for Arnold's, the teen hangout on the show. Locals say the most likely inspiration was the Milky Way, which was located on the North Side, close to a high school, but went out of business twenty years ago.

Two other drive-ins from the 1950s are left in the city, and both offer plenty of neon and 1950s ambiance. Neither, however, has inside seating as at the fictional Arnold's.

### What to See in Milwaukee, Wisconsin

- **Kitt's Frozen Custard,** 7000 West Capital Drive, in the North Side of the city. Eugene Kittredge runs the stand, started by his father and grandfather in the 1950s. The sign on the roof is original porcelain, not more modern plastic, and the neon is

The ambiance of the 1950s can still be found at Kitt's Frozen Custard. *(Photo courtesy of Kitt's Frozen Custard.)*

cream and brown. Frozen custard is made on the premises and is used in such creations as the Hot Fudge Royal Deluxe. Hamburgers and barbecue are also served. The stand is open 10:30 A.M. to midnight, on weekdays, and until 12:30 A.M. on weekends.

▶ **Leon's,** 3131 South 27th Street, in the South Side. Fifties nostalgia is also alive and well at this neon-lit drive-in, built in 1942 and remodeled in 1955 to its current state, according to owner Ron Schneider. The stand offers frozen custard and also has a limited sandwich menu. It is open daily 11 A.M. to midnight (later on weekend nights). Schneider says he does not allow the place to be used as a hangout. But he does have plenty of regular customers.

(See also Los Angeles)

## Fun Facts

• Actor Henry Winkler has a master of fine arts degree from the Yale School of Drama. His performance as Fonzie was so

Leon's Frozen Custard. *(Photo courtesy of the Milwaukee Department of City Development.)*

popular, the character's leather jacket is part of the collection at the Smithsonian (it was donated to the museum in 1980). After *Happy Days*, Winkler's career included co-producing *MacGyver*.

• Ron Howard *(The Andy Griffith Show)* left *Happy Days* in 1980, and later went into directing, with hit movies including *Cocoon*, *Parenthood*, and *The Paper*.

• Series creator Garry Marshall went on to direct hit movies including *Pretty Woman*.

• "Sit on it, Potsie!" was heard from Richie (Ron Howard) when he was angry with his pal Potsie (Anson Williams).

• Actor Anson Williams (Potsie) also went into directing, after first doing a stint in Las Vegas as a lounge singer. He directed episodes of *L.A. Law*, as well as plays in the L.A. area.

• Less successful in her career was Erin Moran (Joanie), who told *People* magazine in 1992 that every member of the *Happy Days* cast was "evil" and that she prays for them. Moran was the only star who did not appear in the show's 1992 reunion special.

• The original theme song for the show was "Rock Around the Clock," by Bill Haley, which had been a hit in 1955 and became a hit again. In 1976, a new original song, "Happy Days," was introduced and also became a hit.

• The cast had an active softball team that played U.S. troops in Germany and Japan on USO tours. Participation on the team by cast members was strongly encouraged by producer Garry Marshall.

## LAVERNE & SHIRLEY
**(January 1976–May, 1983, ABC)**

This popular slapstick comedy was a spin-off of *Happy Days*, and like its predecessor was set in Milwaukee in the 1950s. The characters of Laverne and Shirley, introduced on *Happy Days* as double dates of Richie (Ron Howard) and Fonzie (Henry Winkler), were a bit more down-scale than the *Happy Days* gang. The duo, loud-mouthed Laverne DeFazio (Penny Marshall), she of the big letter "L" on all her clothes, and prim Shirley Feeney (Cindy Williams), work together at the Shotz Brewery, in the bottle cap division, and share a basement apartment in the city. They hang out, but not because they want to, with Squiggy (David L. Lander) and Lenny (Michael McKean), a pair of truck drivers for the brewery.

Laverne's father, Frank (Phil Foster), owns Pizza Bowl, a pizza parlor and bowling alley, where the duo sometimes help out. Frank eventually marries Mrs. Babish (Betty Garrett), Laverne and Shirley's landlady.

There is no Shotz Brewery in real life, but the inspiration is believed to be the old Schlitz Brewery, now closed. Other local breweries are open for tours. Milwaukee landmarks also appeared, or are mentioned, on the show, which was mostly filmed in Hollywood.

In the fall of 1980, in a bid to boost ratings, everyone on the show moved together to Southern California, and the women went to work at a department store there.

While Laverne and Shirley were best friends on the show, the same could not be said of actresses Marshall and Williams. There was discord reported on the set, with Williams claiming that her Shirley character was relegated to a supporting role to Marshall's Laverne, because Marshall and her brother, Garry Marshall, produced the show.

In 1982, Cindy Williams became pregnant and asked the producers for shorter hours, eventually walking off the set when her request was not met to her satisfaction. She later sued the studio, Paramount.

## What to See in Milwaukee, Wisconsin

- **Milwaukee City Hall,** 200 East Wells Street. This building was seen in the show's intro. It is of Flemish Renaissance design, with the first two floors built of granite and the top six floors of brick and terra cotta. There are about 4 million bricks in the building's bell tower, which is named Solomon Juneau, in tribute to the city's founder, a French trader. The building was dedicated in 1896.

- **The basement apartment at 706 E. Juneau.** The first address given on the show was 730 Knapp Street, but in real life there is no apartment building at that address. There is, however, a building a block away at this North Side address that has a high stoop and a basement apartment. This is the place locals thought of when they saw the show, sources at City Hall said. But filming was not done here. It's not far from the old Schlitz Brewery.

- **The Pfister Hotel,** 424 E. Wisconsin Avenue. When Laverne and Shirley went out for a fancy occasion on the show, they sometimes mentioned this luxury landmark hotel. Built in 1893, the hotel has 307 rooms as well as three restaurants, a nightclub, and a lounge. Call 414-273-8222.

- **Miller Brewing Company,** 4251 W. State Street. The city's largest brewery offers free year-round guided tours, every day but Sunday. The brewery is also closed on holidays. A giant-screen video presentation and product samplings are included in the tour, and there's a gift shop selling souvenirs. Call 414-931-BEER.

- **Pabst Brewing Company,** 915 W. Juneau Avenue. One of the oldest breweries in the city, Pabst offers a tour that includes the brewing, packaging, and shipping areas of the brewery. Admission is free. Call 414-223-3709.

Milwaukee City Hall.
*(Photo courtesy of the Milwaukee Department of City Development.)*

▶ **The Historic Schlitz Brewery Park Complex,** 221 W. Galena. The brewery is closed and the building is now an office park, but there is also a nifty restaurant, the Brown Bottle Pub, where one hundred beers from around the world are served. It's open Monday through Friday, from 11 A.M., and Saturday from 5 P.M. Call 414-271-4444.

▶ **Red Carpet Bowling Centers.** This chain offers two locations in the city, the Red Carpet Lanes Celebrity at 5727 S. 27th Street, and the Red Carpet Lanes Regency at 6014 N. 76th Street.

### Fun Facts

• Before working at the brewery, both Laverne and Shirley went through Army basic training.

• Carmine Ragusa (Eddie Mekka) was nicknamed the Big Ragu, and had a big crush on Shirley.

• Ed Begley, Jr. *(St. Elsewhere)* appeared on the show as Shirley's brother, Bobby Feeney.

• Actor David L. Lander also played Squiggy's sister. Squiggy's real name was Andrew.

• Penny Marshall, like her brother, Garry, who produced the show (see also *Happy Days*), became a successful movie director *(Big, A League of Their Own)*.

• The show's theme song, "Making Our Dreams Come True," became a hit for Cyndi Grecco, who also sang the song on the show.

• A cartoon version of *Laverne & Shirley* aired from 1981 to 1983.

• *Laverne & Shirley* was the number one show of the 1977–1978 and 1978–1979 seasons.

## THE MARY TYLER MOORE SHOW
**(September 1970–September 1977, CBS)**

Minnesota's most famous TV landmark is the Victorian house in Minneapolis known to fans of the popular sitcom *The Mary Tyler Moore Show* as the home of much-loved Mary Richards. Other real sites seen on the show include the pedestrian mall area in downtown Minneapolis where Mary confidently threw her hat into the air in the show's intro. And the restaurant table where Mary sat in the intro is also real and now has a plaque noting its place in television history.

On the show, Moore played a single woman in her thirties who was not particularly man crazy and focused more on her career as a television news producer. Mary Richards also had the unique ability to "turn the world on with her smile . . ." She rented an apartment in a house owned by snooty landlady Phyllis (Cloris Leachman), with loud-mouth transplanted New Yorker Rhoda (Valerie Harper) living upstairs.

The house seen as Mary's on the show is located in a wealthy neighborhood of Victorian mansions, near the Lake of the Isles (the lake is also seen in the show's intro).

The house was sold a few years ago for a half million dollars, and a local realtor said the likelihood of finding an apartment like Mary's in the posh neighborhood today is slim.

The original owner of the house owned a farm implement manufacturing company. But the owner at the time of *The Mary Tyler Moore Show* attracted more attention. After giving permission for use of the house's exterior on the show (interiors were filmed on a set in L.A.), she apparently became angry when too many fans showed up outside her residence. So when a film crew came to shoot additional exterior shots in 1974, the house's owner hung an "Impeach Nixon" sign out a window. When she refused to take the sign down, the producers moved Mary, the following season, to a high rise (the move also coincided with Phyllis's departure for a new series, *Phyllis*, based in San Francisco).

The house where Mary Richards lived. *(Photo © 1979,* Star Tribune/*Minneapolis–St. Paul.)*

On the show, Mary's fictional address was 119 N. Weatherly, Apt. D. When she moved to the high rise, it was at 932 N. Weatherly.

## What to See in Minneapolis, Minnesota

- **The house at 2104 Kenwood Parkway.** This private residence was Mary's house on the show. The house was built around 1891 and has been used as both a single family and multi-family home. A third-floor attic apartment was converted to bedrooms by the current owner. A realtor familiar with the property said the house has at least five bedrooms, not including the third floor.

- **Cedar Riverside Plaza,** 1610 6th Street South, in southeastern Minneapolis (across the street from the University of Minnesota West Bank campus). Exterior shots of this high-rise apartment building were used for Mary's second apartment on the show. (Visitors should use caution in the area at night.) While near the campus, visitors may also want to check out the University's funky (yes, that word does apply to some things in this city) Frederick R. Weisman Art Museum on the East Bank, which wasn't there when Mary was but is striking enough to be worth a visit nonetheless.

- **The Crystal Court and Basil's Restaurant at the Marquette Hotel,** 7th Street (between Marquette and Nicollet). The enclosed atrium area of the fifty-one story IDS Center building (the biggest skyscraper between Chicago and L.A.), is where Mary is seen riding the elevator in the show's intro. She is also seen dining at what is now Basil's, a 120-seat restaurant at the Marquette Hotel. The restaurant is on the hotel's third floor and overlooks the Crystal Court atrium area. Visitors can ask to sit at Mary's table on the balcony. For hotel or restaurant reservations call 612-333-4545.

- **The Nicollet Mall Pedestrian Shopping Area,** on 7th Street. This is where Mary throws her hat into the air in the show's intro.

## Fun Facts

• The original concept for the show called for Mary to be divorced.

• Mary Tyler Moore and her husband at the time, Grant Tinker, founded MTM Enterprises, which later went on to produce such acclaimed series as *St. Elsewhere* and *Hill Street Blues*.

• Mary Tyler Moore played Laura Petrie on *The Dick Van Dyke Show*. She also appeared as a nun in the Elvis Presley movie *Change of Habit*.

• *The Mary Tyler Moore Show* won twenty-seven Emmys.

• All the show's main actors—Valerie Harper as Rhoda, Betty White as Sue Ann Nivens, Cloris Leachman as Phyllis, Ted Knight as Ted Baxter, Gavin MacLeod as Murray Slaughter, and Ed Asner as Lou Grant—became television stars.

• One of Mary's best lines was, "I'm an experienced woman. I've been around . . . well, all right, I might not've been around, but I've been . . . nearby."

• Mary works at WJM-TV, channel 12, where she originally applied for a job as a secretary. Her pay as an associate producer at WJM-News was less than she would have made as a secretary. She is promoted to producer in 1975.

• The theme song for the show was "Love Is All Around."

• Mary's parents called her Bones.

• Spin-offs of *The Mary Tyler Moore Show* were sitcoms *Phyllis* and *Rhoda*, and the more serious *Lou Grant*.

• Character Ted's favorite color is plaid, and that's what his wife, Georgette (Georgia Engel), wore to their 1975 wedding.

• Lars (Phyllis's husband) was never seen. He fooled around with horny Sue Ann (Betty White), however.

• Chuckles the Clown was a major star of WJM-TV, but was also never seen. He died in a parade, when, while dressed as a peanut, he was crushed by a hungry elephant.

• Actor Ted Knight's real name was Tadeus Cheslav Wladyslaw Konopka. He died on August 26, 1986.

• First Lady Betty Ford appeared by phone on an episode in 1976. Mrs. Ford calls to talk to Lou Grant and gets Mary on the phone.

## PICKET FENCES
**(September 1992– , CBS)**

This quirky well-acted melodrama, featuring Sheriff Jimmy Brock (Tom Skerritt) and his doctor wife, Jill (Kathy Baker), as the lead characters, deals with a myriad of social issues. Plots for the show are so diverse and creative it has managed to be compared by critics to both *Murder, She Wrote* and *Twin Peaks*, as well as to *The Waltons* and *Dallas*. The show is produced and largely written by David E. Kelley, earlier executive producer/writer on *L.A. Law.*

The small-town setting seen on *Picket Fences* is a Hollywood set. But a real Rome, Wisconsin (one of two in the state), population two hundred, is featured in the show's intro. The still photos in the intro include the fire station in the unincorporated village, which is part of the town of Sullivan, in Jefferson County. There is no Rome police station. The fictional Rome is said by locals to more resemble Jefferson, the seat of Jefferson County, where there is both a city and county police department, as well as a hospital.

### What to See in Rome, Wisconsin

- **The Village of Rome,** in Sullivan (about halfway between Milwaukee and Madison, and about ten minutes from I-94). According to a local resident, the village consists of a general store (see below), a fire station, an antique shop, a few assorted bars, a bowling alley, and "not much else." Actually, the Rome Community Center is also in Rome, and there is a part-time town museum, open on holidays only. The town's population grows in the summer, as there are a lot of nearby campgrounds, and Rome Pond is popular for fishing.

- **Pickets,** Main Street (Highway F). The owner of this general store got wind of the show even before it was on the air and decided to name the store in its honor. Merchandise for sale here includes food and picnic items. Travel guides for Jefferson County are also available. The owner was also toying with the idea of developing *Picket Fences* T-shirts and other items. The store is open daily, 8 A.M. to 8 P.M. Call 414-593-5336.

### Fun Facts

- Kathy Baker, who plays Jill Brock, has an advanced degree from the Cordon Bleu cooking school in Paris.
- Two CBS affiliates dropped the show after an episode included a story line about Mormons and polygamy. One station, in Seattle, reinstated the show quickly, but the other, in Salt Lake City, vowed not to bring it back. Both stations were owned by a subsidiary of the Mormon Church.
- Ray Walston *(My Favorite Martian)* plays the town's judge.
- Producer David E. Kelley married hot, hot, hot movie actress Michelle Pfeiffer in 1993.
- Guest stars have included former New York mayor Ed Koch and CBS newsman Harry Smith. Both appeared on a 1993 episode, in which the town's mayor (Richard Cornthwaite) killed a man who tried to carjack him and who threatened his family. Smith interviewed Koch about his opinions on the Rome case.

## STAR TREK
**(September 1966–September 1969, NBC)**

"... to boldly go where no man has gone before."

"Beam me up, Scotty."

Riverside, Iowa, is the self-proclaimed "future hometown" of Captain Kirk (William Shatner), who is to be born on March 22, 2228. The town gained that designation in 1983, when an enterprising resident, who happened to be on the town council, read that *Star Trek* creator Gene Roddenberry had written the character of Kirk as being born in a small town in Iowa. And Riverside, which has a population of less than 850, decided it would be that town.

Subsequently, a reference to Riverside was made in one of the six *Star Trek* movies, even though the town was never mentioned on the earlier TV series. To celebrate its self-proclaimed distinction, Riverside erected a twenty-foot replica of the starship *Enterprise* in a town park, paid for by the Riverside Area Community Club. The replica was built based on actual blueprints of the starship used on the TV show, provided by Paramount Pictures.

The town also established a *Star Trek* festival called Trek Fest, which is held annually on the last Saturday in June. The festival includes a parade and street dance, and often Trekkies attend in Starfleet uniforms. A swap meet is also held featuring Star Trek memorabilia and other items. Dawn McCoy, secretary for the festival, said of the event, "You don't have to be a Star Trek fan to come, but it helps."

The town also sponsors a birthday party for Kirk on March 22, featuring birthday cake and a punch created by a local bartender, called Romulan ale. Actually, the birthday party has created a bit of controversy among devoted *Star Trek* fans. At the instigation of insistent fans, the town changed the date of the birthday to March 26, which the fans stipulated was the correct date. But the town changed it back to March 22, after discovering that that is the date of actor William Shatner's birthday.

At the town's senior center on Main Street, a sign greeting senior citizens reads "Come Dine with Riverside's Future son, Captain Kirk." And driving into town on Interstate 218 (the All Saints Highway), visitors are greeted with a sign that welcomes them to "Riverside, where the Trek begins." The sign originally said "where the best begins," but years ago someone crossed out "best" and wrote in "Trek" and the sign has remained that way.

Town residents have toyed with the idea of also erecting a Captain Kirk statue, but up to now determined the cost of gaining permission to use the likeness of Kirk (and Shatner) is a bit too costly. "We're just a small farming community and there is no way to raise that much money," said McCoy.

The town promotes its Trekkiness in a county-wide promotion, marketing together with nearby Kalona, an Amish town, as an area where visitors can see the past, present, and future. Money raised from the Star Trek activities goes to the Riverside Area Community Club, a nonprofit organization that puts the money back into the community in the form of community programs, school activities, college scholarships, senior programs, and other community needs. "It's a good way to bring some extra income into town," McCoy said.

## What to See in Riverside, Iowa

- **The starship *Enterprise* replica,** Legion Park, on Highway 22 (off Interstate 218), east of town. This monument to Trekkiedom is known in town as the USS *Riverside.*

- **The welcome to Riverside sign,** visible off Interstate 218. The sign welcomes visitors to the land of Kirk.

- **Star Trek headquarters and Kirk's birthplace,** 50 Main Street. The office for the annual festival sells Star Trek-related souvenirs. Behind the building, which used to be a barber shop, is where Kirk will be born, or so a sign on a stone pillar states. For information on *Star Trek* events in town call 319-648-KIRK.

- **Riverside Flower Trek,** 99 East First Street. This is a flower shop, but *Star Trek* souvenir items are also for sale.

The *Starship Enterprise*™ replica. *(Photo by John Van Allen.)*

Trekkies are greeted in Riverside with this sign. *(Photo by John Van Allen.)*

## FUN FACTS

• *Star Trek* was not originally a successful show. A letter-writing campaign by fans after its second year saved it from oblivion. One million fans wrote letters of protest when the show was canceled after its third season.

• *Star Trek: The Next Generation,* which starred Patrick Stewart as *Enterprise* Captain Jean-Luc Picard, did much better in its first run, lasting seven seasons (October 1987 to May 1994, syndicated). Two spin-offs followed, *Deep Space Nine* in 1993, and *Star Trek: Voyager* in 1995, as well as a movie, *Star Trek: Generations.*

• In his book, *Star Trek Memories,* William Shatner said that appearing on the show was "first and foremost a job," and that he didn't understand what all the fuss from fans was all about.

• Writer Shatner also reveals that actor Lloyd Bridges was first offered the part of Kirk, but turned down the role.

• The show took place in the twenty-third century.

• The *Enterprise* on the show was as large as an ocean liner and ventured through several galaxies. It was commissioned by the United Federation of Planets. The five-year mission of the *Enterprise*, with its crew of four hundred, was to "seek out new life and new civilizations."

- The Romulans and the Klingons were the bad guys.
- Mr. Spock (Leonard Nimoy), Kirk's unemotional sidekick, was a bi-life form, half Earthling and half Vulcan.
- There was a cartoon version of the show that ran from September 1973 to August 1975 on NBC, and featured the voices of William Shatner (Captain James T. Kirk), Leonard Nimoy (Mr. Spock), DeForest Kelley (Dr. Leonard "Bones" McCoy), Nichelle Nichols (Lieutenant Uhura), George Takei (Mr. Sulu), Majel Barrett (Christine Chapel), and James Doohan (Scotty). Walter Koenig (Pavel Chekov) did not work on the cartoon.
- Actress Majel Barrett (Christine Chapel) was married to *Star Trek* creator and executive producer Gene Roddenberry.
- Gene Roddenberry died in 1991, and his ashes reportedly made it into space in 1992, with a real Shuttle astronaut.

## FYI: ALSO IN THE MIDWEST

On *The Beverly Hillbillies* (September 1962 to September 1971, CBS), the suddenly rich hillbilly Clampett family lived in posh Beverly Hills. But when the clan got homesick, it headed to the rolling hills of the Ozarks. And Silver Dollar City, an attraction near Branson, Missouri, was shown as the Clampetts' hometown.

Featured on the show were the 1881 hotel (really an ice cream parlor) and other buildings on the square in Silver Dollar City, which offers a re-creation of life in the late 1800s.

Series creator Paul Henning is from Missouri, and driving from Branson to Silver Dollar City on Highway 76, visitors pass the Ruth and Paul Henning State Forest, which is on land donated to the state by the Hollywood writer/producer.

Henning also chose the Ozarks as the permanent home for the Clampett's famous rundown car, which they ride through the streets of Beverly Hills on the show. It is on permanent display at the Ralph Foster Museum at the College of the Ozarks, also located near Branson.

## What to See in Branson, Missouri

▶ **Silver Dollar City** (off Highway 76, just north of Branson). This theme park, which was seen in six episodes of *The Beverly Hillbillies,* offers a combination of a look at 1890s life and crafts mixed with modern-day amusements, rides, and entertainment, and a real cave that visitors can tour. Staffers dress in period garb and craftspeople demonstrate glassblowing, basket weaving, woodworking and other crafts. Street performers re-enact town scuffles and musicians offer country and western and traditional tunes. The park also offers a free evening music show featuring "Country America's Top 100 Hits." Many craft items are for sale, as are an array of food offerings. Admission is charged. The attraction is open daily from late May to late October, and Wednesday through Sunday in the spring and fall. Hours vary by season. Call 417-338-2611.

▶ **Ralph Foster Museum, College of the Ozarks,** Point Lookout, Missouri. The Clampetts's car from *The Beverly Hillbillies* is on display on the first floor of this museum, where the eclectic collections have earned it the nickname "the Smithsonian of the Ozarks." A cast photo is on display in the car, and visitors can have their picture taken with the car too. The 1921 Oldsmobile was brought on a flatbed truck to the museum from Hollywood. The museum's collection also includes guns, natural history items, and a display on Kewpie dolls. The museum is named for Ralph Foster, a radio pioneer. At the museum gift shop, magnets with *The Beverly Hillbillies* car are among the items for sale. The museum is open 9 A.M. to 4 P.M., Monday through Saturday, and 1 P.M. to 4 P.M. on Sunday. Admission is charged for those over age eighteen. Call 417-334-6411.

For visitor information on Branson, call the Branson/Lakes Area Chamber of Commerce, 417-334-4137.

(See also Beverly Hills and Los Angeles.)

Crazy Corporal Klinger (Jamie Farr) on *M*A*S*H* (September 1972 to September 1983, CBS), loved the hot dogs at Tony Packo's Cafe, in his hometown of Toledo, Ohio, so much he even had them mailed to him in Korea.

But the landmark Toledo, Ohio, eatery isn't the only local attraction mentioned on the show, thanks to Actor Farr, like Klinger, a Toledo native.

Silver Dollar City. *(Photo by Fran Golden.)*

*The Beverly Hillbillies* car, on display at the Ralph Foster Museum, College of the Ozarks, in Point Lookout, Missouri. *(Photo courtesy of Dr. Bobby G. Hendrickson and Ralph Foster Museum, College of the Ozarks, Point Lookout, Missouri.)*

Klinger, who dons women's clothes on *M*A*S*H* in a fruitless attempt to have himself discharged from the Army on mental grounds, is also a booster of the Mudhens, the real Toledo-based Detroit Tigers farm team, for instance.

In real-life, the cast of *M*A*S*H* had Packo's deliver hot dogs to California for cast parties, a Packo's spokeswoman said.

In addition to culinary delights, Packo's is famous for its autographed hot dog bun collection, which includes most of the *M*A*S*H* cast.

The bun-signing tradition started in 1972, when none other than Burt Reynolds *(Evening Shade)* happened into the cafe. Reynolds signed a real bun, but since that time, less perishable styrofoam models have been used.

## WHAT TO SEE IN TOLEDO, OHIO

▶ **Tony Packo's Cafe,** 1902 Front Street. The cafe was founded more than 60 years ago by Tony and Rose Packo, both of Hungarian background. It quickly became known for its Hungarian sausage, which is cut in half and covered in special hot sauce, with mustard and onion, on a soft bun. The cafe also serves other Hungarian delights including chicken paprika and homemade strudel. Both bar and sit-down service is offered. Open Monday through Thursday, 10 A.M. to 11 P.M.; Friday, 10:30 A.M. to 1 A.M.; Saturday, 11 A.M. to 1 A.M. and Sunday, noon to 9 P.M. Call 419-691-6054.

(See also Los Angeles County and Environs.)

The radio sitcom *WKRP in Cincinnati* (September 1978 to September 1982, CBS, then syndicated in the 1990s) featured disc jockeys and other wacky characters at a failing radio station gone rock and roll in Cincinnati. But the show was shot mostly in L.A.

A few Cincinnati sites appeared in the show's intro, however, and nervous newsman Les Nesspin (Richard Sanders) really came to Ohio for one episode in which he landed a biplane, used for his traffic reports, on a road at the King's Island amusement park.

## What to See in Cincinnati, Ohio

▶ **Fountain Square** (center of downtown). Seen in the intro, the Tyler Davidson Fountain and open plaza area is a popular place for office workers to eat lunch outdoors in warm weather.

▶ **The Brent Spence Bridge** (I-75). This bridge, which links Cincinnati to Kentucky, was also seen on the show, according to locals.

## What to See in Mason, Ohio

▶ **Paramount's Kings Island,** 6300 Kings Island Drive (off I-71, north of Cincinnati). Les Nesspin landed a plane at this large theme park, which offers amusement rides and a water park, as well as live plays and performances. Popular rides include the Days of Thunder, which puts riders behind the wheel of a simulated NASCAR racing car. There's also a Top Gun roller coaster, as well as The Beast, the longest wooden coaster in the world. Other shows that have shot on location at the park include *The Partridge Family* and *The Brady Bunch*. Call 513-573-5700.

# CHICAGO

The cop show *Crime Story* and the 1990s version of *The Untouchables,* which featured U.S. Treasury agents fighting the legendary Chicago criminal Al Capone, both came to Chicago seeking a gritty period ambiance. And both shows shot on location, in the real streets of the Windy City.

*Crime Story* went after a 1950s aura and *The Untouchables* a 1930s one, and both productions found plenty of settings to suit their needs, while managing to avoid shots that would indicate the city in more modern times.

With a slightly different task, the camera crews of *Hill Street Blues* tried to avoid showing that the setting was Chicago altogether. Crews from the urban cop show shot exteriors in Chicago, but since the setting for *Hill Street Blues* was supposed to be any major city in the United States east of Chicago, care was taken not to show any recognizable Chicago landmarks (the police station shown is a real one, however).

Sitcoms that have plots based in Chicago have been less prone to such subtleties. Chicago landmarks are sometimes featured to set the scene.

*The Bob Newhart Show* was shot in Studio City, California, but recognizable locations in the Windy City, including the Wabash Bridge, filled the show's intro.

The urban sitcom *Good Times* showed the outside of the real Robert Taylor Homes housing project on State Street in South Chicago, in its intro (an area not advisable for tourists). And *Perfect Strangers* featured shots of Wrigley Field.

Also featuring the famous Chicago ballpark in at least one scene shot on location was *Chicago Hope*. But most of the action on the Chicago-based medical drama really takes place in a studio in L.A. Likewise, *ER*, the hit medical drama also set in Chicago, is filmed mostly in L.A., with crews from the show visiting the Windy City every few months. Sites used have included the University of Illinois at Chicago Medical Center (1740 West Taylor).

*Due South*, the Canadian-made show about a Royal Mountie living in Chicago, isn't really shot here at all, but rather is done on location in Toronto.

According to local sources, *Married . . . With Children*, Fox's Chicago-based adult sitcom, shows the outside of a house in the real suburb of Deerfield as the home of the Bundy family. But no one in town hall seems to know the exact address (it's on Castlewood Lane, somewhere south of Deerfield Road, where the houses kind of look alike, one official explained).

Alas, there is no Lanford, Illinois, the fictional home of the Conner family on the blue-collar sitcom *Roseanne*.

## THE BOB NEWHART SHOW
**(September 1972–September 1978, CBS)**

Bob Newhart's first hit sitcom featured the comedian as Dr. Bob Hartley, a patient, straight-faced Chicago psychologist, who sometimes fumbled in his own life.

Dr. Hartley lived in a high-rise apartment building in the Windy City with his teacher/principal wife, Emily (Suzanne Pleshette), and he worked in a downtown medical building.

The show was filmed before a studio audience in Studio City, California, but the intro and some exterior shots were filmed in Chicago, and featured real Windy City sites. Newhart, for instance, was seen walking across the Wabash Bridge to the El (elevated subway) on his way home from work.

The actor himself is from Chicago, and, after working first as an accountant, got his start in show business on radio in the Windy City. *The Bob Newhart Show* was produced by the same team, Lorenzo Music and Dave Davis, who did the *The Mary Tyler Moore Show.*

## What to See in Chicago, Illinois

- **Buckingham Plaza,** 360 E. Randolph Street (at Lake Shore Drive). This is the apartment building where Bob and Emily live, as seen on the show.

- **Buckingham Memorial Fountain** in Grant Park (downtown between Michigan Avenue and Lake Michigan). This Chicago landmark is among the sites seen on the show. It's modeled after one of the Versailles fountains, and is a popular city rendezvous spot.

- **The Wabash Bridge** (at Wacker Drive and Michigan Avenue). The show's intro has Dr. Hartley (Bob Newhart) walking across this downtown bridge and past such well-known city buildings as the Tribune Tower.

## Fun Facts

• The show was Bob Newhart's second series under the same name. His first was a comedy variety show that aired from October 1961 to June 1962 on NBC. It was not a big hit.

Downtown Chicago, setting for *The Bob Newhart Show.* *(Photo by Peter Schultz, courtesy of the City of Chicago.)*

• Bob Newhart's subsequent shows included *Newhart,* in the 1980s, and *Bob* (also based in Chicago), in the 1990s. Newhart seems to average a show a decade.

• It wasn't until his third sitcom, *Bob,* that Bob Newhart had a fictional offspring, and she was grown-up Trisha (Cynthia Stevenson). In real life, the comedian has four children.

• On *The Bob Newhart Show,* Carol (Marcia Wallace), the receptionist, married a travel agent named Larry Bondurant (Will Mackenzie).

• Dr. Hartley's clients included Mr. Gianelli, played by Noam Pitlik (who later won an Emmy for directing *Barney Miller*). And Penny Marshall *(Laverne & Shirley)* appeared on the premiere episode of the show as Miss Larson, a stewardess.

• Bill Daily, who played neighbor Howard, an airline pilot, played a pilot of a different sort as astronaut Roger Healey on *I Dream of Jeannie.*

## CRIME STORY
**(September 1986–March 1987 and June 1987–May 1988, NBC)**

This cop versus the underworld show, based in the late 1950s, was created by Michael Mann of *Miami Vice* fame, and the aura of the show was dark and often violent.

At the beginning of the series, Lieutenant Mike Torello (Dennis Farina) works for the Chicago Major Crime Unit. But when his main adversary, mobster Ray Luca (Anthony Denison) heads to Las Vegas, Torello follows as part of a special federal strike force. The crime fighter eventually captures his man in a fictional Latin American country.

Joining Torello in his pursuit of Luca is liberal prosecutor David Abrams (Stephen Lang), who later heads the strike force.

The show was shot on location. In Chicago, the producers sought settings that weren't much changed from the 1950s, according to Ron Verkuilen, field scout for the Illinois Film Office. The period locations were shot close up, so as not to catch more modern buildings or cars in the background.

A lot of filming for *Crime Story* took place in the Wicker Park neighborhood, and the producers went to the Chicago suburb of Park Ridge for the setting of Ray Luca's home. A number of local eating establishments were also featured. As the show's plot moved, location shooting followed, from Chicago to Las Vegas, and later to Acapulco.

## What to See in Chicago, Illinois

- **Superdawg,** at Milwaukee, Devon, and Nagle streets (in the northwestern corner of the city, about ten minutes' drive from O'Hare Airport). This landmark carhop operation has been around since 1948, serving Superdawgs, which owner Maurice Berman stresses are not hot dogs or frankfurters or even wieners, but rather their own form of pure beef, made from a special recipe. The stand is where the *Crime Story* characters were seen standing in the opening credits. It also appeared in the movie *Sixteen Candles* and in a long-running Chevy truck commercial. Open daily, 11 A.M. to 1 A.M., and until 2 A.M. on Friday and Saturday. Call 312-763-0660.

Not a hot dog, but a Superdawg, as featured on *Crime Story*. *(Photo courtesy of Don Drucker, Superdawg Drive-In, Inc.)*

▶ **The 12th District, Chicago Police Department Building,** 100 South Racine (at Monroe and Racine, in the Pilson neighborhood). This real precinct house appeared as a police station on the show as well.

▶ **The Busy Bee Diner,** 1546 North Damen Avenue (west of downtown, in the Wicker Park area). Exteriors of this classic diner, which has been in this location since 1913, were used on the show. The Busy Bee is open daily, 6 A.M. to 8 P.M., serving breakfast, lunch, and dinner. Call 312-772-4433.

▶ **The Diner,** 1635 West Irving Park (in the Lakeview neighborhood). Filming for *Crime Story* took place inside this twenty-four-hour diner, with its 1940s decor. The structure is made from real streetcars, and has been open since 1937. Ham and eggs and homemade chili are specialties here. Call 312-248-2030.

▶ **Meigs Field,** 15th Street at Lake Michigan. The show filmed a number of times at this small airport, close to downtown. The field is used by commuter air carriers and for private flights.

(See also Arizona and Nevada)

## Fun Facts

• Before becoming an actor, Dennis Farina was a real-life cop in Chicago for eighteen years.

• The theme song for the show was "Runaway," sung by Del Shannon. Rocker Todd Rundgren did the show's score.

• David Soul *(Here Come the Brides, Starsky and Hutch)* guest-starred in a 1987 episode as the new husband of Torello's ex-wife, Julie (Darlanne Fluegel).

• In the pilot episode, David Caruso *(NYPD Blue)* played a young hood, Johnny O'Donnell, son of Torello's old friends. He was killed off, but not before receiving rave reviews.

• Debbie Harry, formerly of the rock group Blondie, played a girlfriend of Luca, Bambi, in an episode that also featured singer Paul Anka as Luca's courier, Tony Dio.

• Rude stand-up comedian Andrew Dice Clay played Max Goldman on the show.

• In a creative bit of casting (in terms of publicity), five popular disc jockeys were brought in for the show's three-part finale. The DJs from San Francisco, Boston, Chicago, Dallas, and Philadelphia appeared playing various roles not related to their on-air duties.

• In one episode, Luca and his sidekick, Pauli (John Santucci) survived a nuclear bomb blast in the dessert.

## HILL STREET BLUES
### (January 1981–May 1987, NBC)

This intelligent cop show took place in an unnamed eastern city, with the Hill Street Station located in a grim and dangerous inner-city neighborhood. The work and personal lives of the cops and others in the precinct were featured in the story lines. NBC described the unusual show as a "humorous police drama series."

*Hill Street Blues* was shot in Studio City, with synthetic snow used in some winter scenes to establish an eastern setting. But outside scenes of Chicago were also featured. The station house seen on the show is a real Windy City police station. According to Ron Verkuilen, field scout for the Illinois Film Office, the real-life Maxwell Street Flea Market (which moved in 1994, from Roosevelt and 16th to Canal Street) appeared on the show, as well as shots showing warehouses and general big-city grittiness. Train shots were filmed off the Chicago Merchandise Mart's roof (222 West Bank Drive).

*Hill Street Blues* crews carefully filmed so as not to catch Chicago's landmarks in the background, such as the Sears Tower, since the setting was supposed to be any U.S. city east of Chicago.

The show's writers did some research on police issues by talking with representatives of the real Chicago Police Department, Verkuilen said. Urban law enforcement officials applauded *Hill Street Blues* as being realistic, despite characters like Detective Belker (Bruce Weitz), who was known occasionally to bite and snarl at the bad guys, whom he called "dog breath."

*Hill Street Blues* won awards and accolades but never attracted large audiences, despite the fact its witty writing was compared to that of the highly popular show *M*A*S*H*.

## What to See in Chicago, Illinois

▶ **The Maxwell Street Station,** 14th and Morgan streets. Exterior shots of this older police station, with a limestone archway, were used on the show. The building in real life houses an investigative unit of the Chicago Police Department.

## Fun Facts

- Public defender Joyce Davenport (Veronica Hamel) called her lover (later her husband) Captain Frank Furillo (Daniel J. Travanti) "Pizza Man," because he was Italian.
- Michael Conrad, who played Sergeant Phil Esterhaus, the kindly officer who always ended roll call with "Let's be careful out there," died in 1983. On the show his character was written out with the explanation that Esterhaus had a heart attack while making love to ravenous Grace Gardner (Barbara Babcock).

The Maxwell Street Station. *(Photo © 1994, Ron Schramm.)*

• Barbara Bosson, who played Furillo's needy ex-wife, Fay, is married to the series's cocreator/producer Steven Bochco.

• *Beverly Hills Buntz* was a spin-off of *Hill Street Blues*, but it lasted only one season. The show starred Dennis Franz *(NYPD Blue)*. Franz had actually appeared on *Hill Street Blues* in two roles, as Detective Sal Benedetto (who was killed off), then as Lieutenant Norman Buntz.

• Actor Peter Jurasik, who played slimy informant Sidney Thurston on *Hill Street Blues* also took the character to *Beverly Hills Buntz*.

• Officer Joe Coffey (Ed Marinaro) was killed off in 1986, when the actor left the series.

• Officer Andy Renko (Charles Haid) was supposed to be killed off in the pilot, but the character proved to be popular enough to keep around. Actress Debi Richter, who played Renko's wife, Daryl Ann Renko, is married to Haid in real life too. Haid later went into directing; his credits include episodes of *ER*.

• The show's memorable theme song, written by Mike Post, made it to the record charts.

## FYI: ALSO IN CHICAGO

*The Untouchables* (1992 to 1994, syndicated), the second crime series under that title (the first was in the early 1960s), was filmed on location in Chicago, using neighborhoods with a 1930s or earlier ambiance, including the historic area of Pullman.

Production of the show was also done at Chicago Studio City (5660 West Taylor). Exteriors of old warehouses in that neighborhood were also used.

### What to See in Chicago

▶ **Pullman** (near Lake Calumet, on Chicago's South Side). This historic area was built as a model company town by George Pullman, the Chicago industrialist. Seeking Utopia, Pullman created the town on undeveloped prairie land around his Pullman Palace Car Company, where Pullman railroad cars were constructed.

# THE WEST AND THE OLD WEST

Just like on *Dallas,* oil men really do exist in the Texas city, and some can be found in Dallas's big skyscrapers and posh restaurants that appear on the show. There have even been real Ewings in the oil business in Dallas (although not quite as big-time as J.R.).

Outside the city, Southfork Ranch, the large white house made famous in the opening of *Dallas,* is open to the show's fans. It's now a convention center and tourist attraction.

*Mork & Mindy* fans won't say "Shazbat!" (that's Orkan for "Drats!") in Boulder, Colorado, where the real house that appears as Mindy's abode on the show can still be found, along with several other featured sites. The producers chose Boulder for the spacey sitcom because, they said, it seemed likely an alien would end up in the college town.

*Dynasty* made equal sense for Denver, Colorado, where there are some wealthy citizens, except that producers of that glitzy soap said they couldn't find a suitable mansion for the Carrington clan in the Mile High City. So they used a northern California mansion instead, appeasing spurned Denver with a few scenic exterior shots. The show was really shot mostly in Hollywood.

*Perry Mason,* after it was brought back to life in the 1980s, showed a bit more of Denver in episodes shot on location in the city. And *Father Dowling Mysteries* spent some time on location in Denver, too.

The Old West is alive and well in Kansas, where the real Dodge City is full of reminders of *Gunsmoke* (geared toward a tourist crowd). The famous Boot Hill Cemetery was re-created on the show's set, but *Gunsmoke* never actually went on location in Dodge City. The show did go on location to Kanab, Utah, however.

The Beehive State is, in fact, an unlikely gold mine of western TV locations. *The Lone Ranger* and *Wagon Train* are among other shows that shot scenes in Kanab, Utah. And Moab, Utah, is also a popular shooting site.

But Utah hasn't just doubled for other western states. The pilot for *MacGyver* was shot at Dead Horse Point State Park, near Moab, Utah; on the spy show, the location was supposed to be Asia. And according to local sources, on the *MacGyver* episode, members of the Mongolian Army were played by Navajo Indians.

## ALIAS SMITH AND JONES
**(January 1971–January 1973, ABC)**

This western adventure show featured two ex-outlaw characters trying to go clean and hoping to be pardoned by the governor.

Hannibal Heyes (Peter Deuel, then Roger Davis) adopted the alias Joshua Smith, and his partner, Jed "Kid" Curry (Ben Murphy), the alias Thaddeus Jones. The show was set in the 1890s, and the pair were supposed to be from Kansas.

But the pilot and parts of several episodes were shot on location using the canyon areas, rock formations, and other scenery in and around Moab, Utah. Moab, which is in southeastern Utah, has the oldest film commission in the world, with filming taking place steadily in the Canyonlands since 1949.

### WHAT TO SEE IN MOAB, UTAH

- **Castle Valley,** Highway 128 (about sixteen miles northeast of Moab). The Castle Rock and Priest and Nun formations appeared on *Alias Smith and Jones* as well as in other TV shows, movies, and commercials.

- **Professor Valley** (off scenic Highway 128, about twenty-three miles northeast of Moab). This area also appeared on the show, as did the La Sal Mountains and the Potash Road, near Moab.

- **Moab Film Commission,** 50 East Center Street (near the Moab Visitors Center). The commission has a small museum with newspaper clippings, scripts, and photos of TV shows and movies shot in the area.

- **The Moab Visitors Center,** 805 North Main Street. The center offers information on the region including a driving tour of film locations.

- **The Hollywood Stuntmen's Hall of Fame,** 111 East 100 North. This one-of-a-kind museum is a must-see for fans of westerns and other action shows. It's run by John Hagner, himself a longtime stuntman (his TV appearances include doubling for Gardner McKay in the 1960s series *Adventures in Paradise*). The collections at the museum include stunt-related exhibits and two hundred footprints of famous actors. There is a special exhibit on *The Lone Ranger* (the museum has Clayton Moore's footprints). Also on display are rare outtakes from TV shows and movies. Hagner does occasional performances of stunts (just ask him) and is the resident authority on anything involving the tricks of his trade. Also on display are Hagner's own portraits of the stars, copies of which are for sale, along with other souvenirs. Admission is charged. Call 801-259-6100.

### Fun Facts

- Actor Peter Deuel, who was the first actor to play Smith, committed suicide in late 1971. He appeared in the few episodes of *Alias Smith and Jones* that had already been filmed, and was replaced by Roger Davis. Deuel had earlier appeared on *Gidget*.
- Among Ben Murphy's later roles was that of Warren, the son of Pug Henry (Robert Mitchum) and Rhoda Henry (Polly Bergen) in the 1983 miniseries *The Winds of War.*
- Roger Davis also later acted in miniseries land, playing Max Kendrick in *The Innocent and the Damned.*

## DALLAS
**(April 1978–May 1991, CBS)**

All of the city of Dallas was literally a set for this big nighttime soap. The city's reputation as being home to cowboys and oil tycoons was enhanced by the show, and Dallas itself became a star, second only perhaps to J.R. (Larry Hagman), the man everyone loved to hate.

Nanette Farlow, the show's location manager, quoted producer Leonard Katzman as saying the show "turned Dallas from people thinking that's where President Kennedy was shot, to that's where J.R. Ewing lived."

There wasn't much ground in the city left uncovered on the show. During it's run, the cast and crew came from Hollywood to Dallas once every TV season for three months of filming.

The most highlighted setting of the show was, of course, Southfork, a real-life ranch in Parker, Texas, about twenty miles northeast of Dallas (on *Dallas* it was in fictional Braddock, Texas), which is now open to visitors as a convention center and tourist attraction.

On a clear day you can see the ranch when you fly into Dallas from the east. And officials of the ranch say proudly that it is "The most famous white house west of D.C." Hundreds of thousands of visitors stop by the six-bedroom house each year, making it Dallas's most popular tourist attraction.

On the show, the oilmen worked downtown, and several real downtown Dallas buildings were seen as Ewing Oil Company headquarters, with construction projects one reason for the moves (the producers didn't want to show skyscrapers with cranes and scaffolding).

The women on the show shopped at the real upscale mall, The Galleria, and the Ewings dined at real top Dallas spots like the French Room of The Adolphus Hotel.

The stockyards of Forth Worth were seen on *Dallas* too, but the production crew didn't particularly enjoy working there, according to Farlow. The cowboys were rowdy and sometimes started fights not called for in the scripts, Farlow said.

When they were sick, the characters on the show went to Baylor Hospital or Parkland Memorial, where J.R., like President John F. Kennedy, was taken when he was shot. The show also used as a setting the real Love Field and Dallas-Fort Worth International airports.

One time at DFW, during a scene in which J.R. is supposed to be picking up a bag, actor Hagman was accosted by a red-haired woman who had just gotten off a plane from England.

"She went through security yelling, 'You bastard. How could

you treat Sue Ellen that way. I'm going to get you.' And Hagman said to her, 'Truly, truly, I'm not a bastard,'" Farlow said.

The show also occasionally caused controversy among local residents. For instance, a real-life man named Bobby Ewing, who owned an oil firm and lived in Dallas, sued Lorimar Productions after the show's production company tried to prevent him from selling T-shirts, hats, jeans, and paperweights with the name Ewing Oil. An out-of-court settlement was reached, and the real Ewing agreed to stop selling the items.

In addition to Dallas, the show also went on location in Europe.

### What to See in Parker, Texas

▶ **Southfork Ranch and Conference Center,** 3700 Hogge Road (Central Expressway, to Exit 30). Guided tours are offered at the forty-one-acre ranch made famous on the show. There are also facilities available for conferences and events, including a

The famous Southfork Ranch. *(Photo courtesy of Southfork Ranch and Conference Center.)*

63,000-square-foot conference center and the Oil Baron's Ballroom. The interior of the main house has been renovated in a grand style designed to please visitors; plush, but not exactly as on the show. A museum offering a history of the series is in the works; it will include a display of "the gun that shot J.R." Other facilities at the ranch include Miss Ellie's Deli, which offers Texas BBQ and displays photos of the stars; and Lincolns & Longhorns, which features the original Lincoln Continental Jock Ewing (played by Jim Davis, who died in 1981) drove on the show, and which offers western apparel and gift items for sale. Real Texas longhorns graze on the grounds of Southfork. Admission is charged. Call 800-989-7800. Southfork is about a forty-minute drive from Dallas, depending on traffic. For those without a car, the ranch is featured on bus tours by Greyline (214-824-2424) and Longhorn (214-228-4571).

### Fun Feature: Weddings at Southfork

On the show, the ranch was host to several nuptials including those of J.R. (Larry Hagman) to Sue Ellen (Linda Gray), J.R. to Cally (Cathy Podewell), Lucy (Charlene Tilton) to Mitch Cooper (Leigh McCloskey), and Bobby (Patrick Duffy) to Pam (Victoria Principal).

For those who want to follow in the footsteps of the show's characters, Southfork Ranch offers chic weddings through Far Away Weddings, a local wedding consulting firm. The weddings can be held right on the ranch's manicured lawns, complete with western-style hats and music, such as the theme song from *Dallas*, or that of *Bonanza* (apparently frequently requested).

The setting inspires creativity. One couple hired the helicopter firm that was used to fly the *Dallas* production crew around for shoots, so they could make a grand entrance. Another arrived on horseback. And another staged a shotgun wedding. J.R. look-a-likes are available to walk the bride down the aisle.

The ranch also offers honeymooners and other visitors a chance to really experience Southfork by spending the night. But the offer is not for the faint-of-pocketbook. For $3,500 a night, up to eight guests can stay at the mansion. Included in the cost is the attention of a personal wait staff, use of the pool, a Texas-style dinner, a take-home gift of a crystal decanter with eight wine glasses etched with the Southfork logo, and monogrammed his or her robes for the couple who sleep in J.R.'s master suite. Call 214-442-7800.

## A Brief History of Southfork

The fictional history of the ranch is that it was founded in 1858 by Enoch Southworth. Jock Ewing (Jim Davis) married Eleanor Southworth (Barbara Bel Geddes, Donna Reed) and took over when her father, Aaron, died in the 1930s (and Miss Ellie inherited the land).

NationsBank Plaza, one of the real Dallas buildings seen as the headquarters of Ewing Oil on *Dallas*. *(Photo courtesy of the Dallas Convention and Visitors Bureau.)*

In real life, at the time of *Dallas*, the ranch, which was then known as Duncan Farms, was owned by J.R. Duncan (Joe R.), a millionaire land developer, whose family was in residence at the time of filming. The Duncan's built the house in 1970.

Duncan renamed the ranch Southfork in honor of the show, and started selling T-shirts bearing the name. He sold the house in 1984, for $7 million, to Colin Commodore Ltd., to be operated as a hotel. But that venture went out of business.

In 1992, the ranch was purchased at an auction for $2.6 million by Arizona businessman Rex Maughan, chairman and president of Forever Living Products International, the largest grower of aloe vera in the world.

## What to See in Dallas, Texas

▶ **Ewing Oil.** The first building seen as the headquarters for Ewing Oil was the **NationsBank Building** (at Cedar Springs and Turtle Creek). The second was the **Renaissance Tower** (at 1201 Elm Street). And the third was **NationsBank Plaza** (at 901 Main Street).

▶ **Trammel Crow Center,** 2200 Ross Avenue. This was the location used for Sue Ellen's design firm. Her lingerie line was called Valentine. Interiors were shot here too.

▶ **Parkland Memorial Hospital,** 5201 Harry Hines. This was where J.R. was taken after he was shot. It is also where President John F. Kennedy died. Interior and exterior shots of the hospital appeared on *Dallas*.

▶ **Fountain Place,** 1445 Ross Avenue. Bobby Ewing's office was here when he wasn't working with J.R., at Ewing Oil.

▶ **Doubletree Towers,** 8250 North Central Expressway (across from the NorthPark Mall). Cliff Barnes worked out of offices here.

▶ **Sheraton Park Central Hotel & Towers,** 12720 Merit Drive, off the LBJ Freeway (Highway 635). The hotel appeared as itself on the show. Some of the cast and crew stayed here during filming in Dallas. Call 214-385-3000.

▶ **The Mansion on Turtle Creek,** 2821 Turtle Creek Boulevard. This luxury hotel is where many of the stars of the show, including Linda Gray, Larry Hagman, and Victoria Principal, stayed while in Dallas. Call 214-559-2100.

▶ **The Galleria,** 13344 Noel (where the North Dallas Toll Road and the LBJ Freeway, Highway 635, intersect). This was where Sue Ellen and the other women of *Dallas* shopped for furs and other baubles.

▶ **The Palm,** 1321 Commerce Street. This downtown bar appeared frequently on *Dallas*, mostly because filming here was allowed for free. Call 214-742-8200.

▶ **The West End.** This historic area was where the fraudulent Jock (C. Forest) courted Miss Ellie (Barbara Bel Geddes) on a surrey ride.

▶ **Stouffer Dallas Hotel,** 2222 Stemmons Freeway. According to location manager Farlow, the top of this thirty-story hotel offers the best view of downtown. It was used for shots of downtown seen on *Dallas* several times. Call 214-631-2222.

▶ **The Adolphus Hotel,** 1321 Commerce Street. This grand hotel was used quite a bit on the show. The renowned French Room restaurant at The Adolphus was also featured. Call 214-742-8200.

For more information on the city, call **The Dallas Convention & Visitors Bureau** Special Events Hotline, 214-746-6679. Or visit the visitor center at Union Station.

## What to See in Celina, Texas

▶ **Punk Carter Cutting Horse,** Route 2 (about thirty miles north of Dallas). This working ranch was used for cow and horse shots on *Dallas*. The main business here really is training horses, but the ranch has also gotten into hosting "City Slicker" parties, offering cookouts, country music, cowboy poetry, roping, and branding as in the movie *City Slickers*. The ranch also has a cabin with scenic views, available for nightly rental. Call 214-382-2849.

## What to See in Fort Worth, Texas

▶ **The Stockyards,** North Fort Worth (about two miles from I-35W). In the late 1800s, this area was one of the primary cattle receiving areas in the Southwest. Today it is an historic area and tourist attraction, with strolling cowboys and the famous Billy Bob's honky-tonk among venues providing entertainment. There's also a rodeo at the coliseum most weekends, and visitors can pose for pictures seated on a longhorn steer. A visitor center on western-looking Exchange Street provides information on current attractions. Call 817-624-4741.

## What to See in Justin, Texas

▶ **Justin Discount Boots and Cowboy Outfitters,** Highway 156 (about one hour from Dallas, or thirty minutes from Fort Worth). It's a bit of a haul, but locals say one of the best places in the area to buy a ten-gallon hat or other cowboy supplies is this small-town store. Hats, shirts, belts, and boots are among the items for sale. The store is open Monday through Saturday, 9 A.M. to 6 P.M. Call 817-430-8084.

## Fun Facts

• On November 21, 1980, 41.5 million people tuned in to see who shot J.R. (Larry Hagman). It was Kristin Shepard (played by Bing Crosby's daughter, Mary) who shot J.R., but she was not prosecuted. She did have to leave town, however.

• *Dallas* was shown in more than ninety countries. The show was so popular in Israel, that it was broadcast on army radio. Theater owners in Johannesburg, South Africa, closed on days the show was aired, rather than compete with *Dallas* for patrons. The show immediately flopped in Japan, however.

• Jackie, initially Cliff's secretary and later a Ewing employee, was played by series producer Leonard Katzman's daughter, Sherril Lynn Rettino.

• The nemesis of the Ewing family was Digger Barnes (David Wayne, then Keenan Wynn), father of Cliff (Ken Kercheval). Digger's real first name was Willard.

• The part of Jenna was played by three actresses, Morgan Fairchild, Francine Tucker, and then Priscilla Presley.

• Bobby (Patrick Duffy) was elected a Texas state senator on the show in 1980.

• J.R. and Bobby's brother Gary (David Ackroyd, then Ted Shackelford), the middle son, was seldom seen on *Dallas*, but was a lead character in the spin-off series *Knots Landing*. Ray Krebbs (Steve Kanaly), a ranch hand, was the Ewing brothers' half-brother, and Jock's illegitimate son.

• When the character Jock (Jim Davis) died on the show, a mention of a Jock Ewing Memorial Scholarship at Southern Methodist University drew so many inquiries from potential donors that a real scholarship was established for television, film, and communication students.

• Barbara Eden, who costarred with Hagman in *I Dream of Jeannie*, guest-starred on *Dallas* as LeeAnn De La Vega, who buys Ewing Oil.

• Larry Hagman once said of his portrayal of J.R., "I just do it for the money."

• J.R.'s favorite drink was bourbon and branch water, but in real life, actor Hagman was said to favor beer, champagne, and vodka.

## DYNASTY
**(January 1981–May 1989, ABC)**

*Dynasty* was to be Denver's *Dallas*, but actually most of the action for the nighttime soap opera was shot in California, and not on location.

The producers received some criticism in Denver, in fact, for not showing more of the city and surrounding landscape (they even had the audacity to show snow-capped California mountains and pretend they were in Colorado). Winter shots of downtown Denver from City Park, showing Colorado's real mountains, were added to appease the locals.

The producers also said they couldn't find a suitable mansion in Denver for the wealthy Carrington clan, so they chose a mansion in northern California instead. Denverites were naturally insulted by the choice. And a local newspaper columnist in Denver even jokingly offered his own home for the show.

A few episodes were eventually shot on location in Denver,

including one in 1983 in which former President Gerald Ford and Secretary of State Henry Kissinger made a guest appearance.

On the show, the notables attended the real Denver Carousel Ball, a major fund-raising event for the Children's Diabetes Foundation, which was held at Currigan Hall near the convention center. The glittery biannual event was moved to L.A. in 1985. The ball shown on the show was hosted by then Twentieth Century-Fox owner Martin Davis, who, with his wife, still chairs the event.

*Dynasty* was the second big oil soap after *Dallas*, and was purposely campy. Or at least it appeared so. How else to explain Alexis (Joan Collins)? The show stressed "life-styles of the rich and famous," with jewels and fancy fashions galore. The clothing budget for the show was $10,000 a week, at the time larger than that of any other series.

### What to See in Denver, Colorado

- ▶ **City Park,** off Colorado Boulevard. This public park is the largest park in the city (the Central Park of Denver) and is home to the Denver Zoo, City Park Golf Course, and the Denver Museum of Natural History. It's also a great place to see the Denver skyline and nearby mountains, and shots for the show were done here, local sources said.

### Fun Facts

- A variety of products were introduced bearing the Dynasty name, including furs, perfume, tuxedos, women's suits, evening dresses, and even panty hose. In 1985 alone, some $150 million worth of *Dynasty*-related products were sold.
- Yves St. Laurent created the popular Forever Krystle perfume, as well as a men's cologne, Carrington.
- In her first major television role, Heather Locklear played the devious, social-climbing niece of Krystle, Sammy Jo Dean. *Dynasty* brought Locklear to the attention of producer Aaron Spelling, leading to her most popular conniving role, that of Amanda on *Melrose Place*.
- Joan Collins was accused of overplaying megalomaniacal Alexis, but a lot of babies born in the 1980s were named after

A view of Denver, as glimpsed on *Dynasty*. *(Photo courtesy of the Denver Metro Convention & Visitors Bureau.)*

her character. Collins did not join the cast of *Dynasty* until the show's second season.

• Krystle (Linda Evans) was Blake Carrington's (John Forsythe) secretary before they got married. Evans's earlier TV appearances included the part of Audra on *The Big Valley*.

• Linda Evans is an ex-wife of John Derek, later Bo's husband.

• John Forsythe earlier had very different roles in *Bachelor Father*, the story of a lawyer raising his thirteen-year-old niece, and *To Rome with Love*, in which he played a widowed professor with three daughters.

• Two actors, Al Corley and then Jack Coleman, played Steven, the bisexual son of Alexis and Blake. Steven underwent plastic surgery after being burned in an oil rig explosion, which explained the change in actors.

• The part of Fallon, daughter of Blake and Alexis, was

played by two actresses, Pamela Sue Martin and then Emma Samms.

- Rock Hudson made his final TV appearance on *Dynasty*, as Daniel Reece. He later died of complications from AIDS.
- George Peppard (who later starred in *The A-Team*), played the part of Blake in the pilot for *Dynasty*.
- The show spawned a less-successful spin-off, *The Colbys*.

## GUNSMOKE
**(September 1955–September 1975, CBS)**

The longest-running TV western, *Gunsmoke*, played up Dodge City's image of being one of the wildest places in the Wild West.

Marshal Matt Dillon (James Arness), the fictional no-nonsense lawman of *Gunsmoke*, fights for justice in the early 1870s; the character was said to be a composite of several real nineteenth-century lawmen of Dodge City, including Wyatt Earp and Bat Masterson.

In real life, Dodge City did not have as many gunfights as its reputation would imply (the show had at least one per episode). But the city, which was founded in 1872 for buffalo trade, and which later was a major center for cattle trading, was not without its incidents either. During Dodge City's first few years, there was no law in town. And more than one argument was settled with guns. Later it was the bad guys shooting the good guys, and vice versa, with residents including, in addition to Earp and Masterson, Doc Holliday and Ben Thompson.

Boot Hill Cemetery, mentioned on *Gunsmoke*, was created as a burying ground for unsuccessful, unidentifiable gunfighters, and was thus used until 1878. (More respectable citizens were buried at nearby Fort Dodge).

In addition to the rough cowboys, gamblers and prostitutes were attracted to the Santa Fe Trail town, and according to local historians, by 1876 the population of Dodge City was 1,200, with nineteen establishments selling liquor.

The real Dodge City had two Front streets facing the railroad tracks. On the north side of the tracks, guns were not allowed to be worn or carried. The south side was the wrong side of the tracks, and guns were common.

The Dodge City seen on *Gunsmoke* was a Hollywood set, but some aspects of the fictional version were based on fact. In addition to Boot Hill, there really was a Long Branch Saloon in Dodge City, for instance, but presumably without Miss Kitty (Amanda Blake).

Today in Dodge City, *Gunsmoke* fans can find a re-creation of the Long Branch that combines aspects of the real and fictional saloons. It's located at the Boot Hill Museum, also known as Old Dodge City, a living history museum built at the site of the Boot Hill Cemetery (the bodies of the unknown gunfighters buried at the cemetery have been moved elsewhere).

While *Gunsmoke* did not go on location in Dodge City, the actors were known to visit on occasion. A street in downtown Dodge City is named Gunsmoke Street, and several businesses in the city bear the name of the show.

The *Gunsmoke* crew did go on location in other western-looking locales, including Kanab, Utah.

Interestingly, the producers of *Gunsmoke* chose to depict, in their version of Dodge City, a setting with mountains in the background. The real Dodge City does not have mountains in its landscape.

## What to See in Dodge City, Kansas

▶ **The Boot Hill Museum,** Front Street. The museum's several buildings offer a look at what the city was like in the 1870s and 1880s. The buildings were re-created based on pictures and photographs from the period. Included in the museum's collection is an exhibit on *Gunsmoke*, with a dress worn by Miss Kitty (Amanda Blake), other memorabilia, and photos. The museum is open daily in the summer, 8 A.M. to 8 P.M. From September through May, it is open Monday through Saturday, 9 A.M. to 5 P.M., and Sunday, 1 P.M. to 5 P.M. Admission is charged. Call 316-227-8188.

▶ **The Long Branch Variety Show,** at the Boot Hill Museum. The show stars Miss Kitty and her cancan dancers, and opens with a gunfight. It is offered nightly in the summer, at 7:30 P.M. A fee is charged, and reservations are requested. Other entertainment at Boot Hill includes gunfights on the street, stagecoach rides, and medicine shows. Chuckwagon BBQ dinners are also offered nightly. Call 316-227-8188.

The Boot Hill Museum. *(Photo courtesy of the Dodge City Convention and Visitors Bureau.)*

▶ **The Original Site of Front Street,** Wyatt Earp Boulevard (near 2nd Avenue). This is the area where the original Long Branch Saloon and other Front Street businesses stood. Most of the original wooden buildings on Front Street were destroyed by fire.

▶ **Fort Dodge,** Highway 154 (five miles east of Dodge City). Before there was Dodge City, there was a military outpost here. The fort was built in 1865 to protect wagon trains and the U. S. Mail. Several of the fort's original buildings remain. Civilians buried at Fort Dodge Cemetery include Marshal Ed Masterson, brother of Bat.

▶ **The Dodge City Convention and Visitors Bureau,** 4th and Spruce. The bureau offers a cassette tape with an audio narrative of Dodge City history. Call 316-225-8186.

## Fun Feature: Dodge City–Speak

According to the Kansas Heritage Center, five words or expressions are believed to have originated in Dodge City:

A "Miss Kitty" at the Long Branch Saloon, at The Boot Hill Museum. *(Photo courtesy of the Dodge City Convention and Visitors Bureau.)*

Stinker: Described the smell of buffalo hunters
Joint: Used as another name for saloon
Cooler: Was the nickname given to the city's first jail
Stiff: Was used to describe dead bodies
Red Light District: Came from the railroad men who would leave their lanterns outside the city's brothels

## What to See in Kanab, Utah

▶ **Johnson Canyon,** Highway 89 (about ten miles east of Kanab). This canyon area has an old western movie town set. The grounds are private, but you can catch a glimpse along the scenic highway. The area has many caves and rock formations, as well as Indian petroglyphs. In addition to *Gunsmoke*, episodes of several shows were filmed in the area, including *Death Valley Days*, *Have Gun Will Travel*, *How the West Was Won*, *Wagon Train*, and *The Six Million Dollar Man*. Several movies also used the site.

▶ **Paria Canyon Wilderness Area,** Highway 89 (thirty-four miles east of Kanab). *Gunsmoke* was among the shows and movies that filmed at the Paria movie set, now owned by the state and open to the public. To get to the movie set area, you need to ride on a six-mile winding dirt road, marked as a Scenic Backway. The area also offers spectacular red rock formations, and hiking along the Paria River. Episodes of *Death Valley Days* and *Wagon Train* were also filmed here. The public Paria Movie Set Campground is operated by the Bureau of Land Management. For camping and hiking information call 801-826-4291.

## Fun Facts

- *Gunsmoke* premiered on radio in 1952, with William Conrad playing Matt Dillon. He would later star as Detective Frank Cannon on the 1970s detective show *Cannon.*
- John Wayne was the first choice for the TV role of Marshal Dillon. The Duke declined but agreed to introduce the first episode of the show after the producers heeded his advice and hired James Arness for the part.
- James Arness is six feet, seven inches tall, and is the brother of Peter Graves *(Mission Impossible)*
- Reruns of the early half-hour episodes of *Gunsmoke* were rebroadcast under the title *Marshal Dillon.* The show was shown in one-hour form beginning in 1961.
- Burt Reynolds *(Evening Shade)* joined the cast in 1962 as Quint Asper, a blacksmith who was half Indian.
- Dennis Weaver *(McCloud)* played Chester Goode, the deputy.
- In addition to Arness, Milburn Stone, who played Doc Galen Adams, was the only other cast member to be on the show its entire run.

## THE LIFE AND TIMES OF GRIZZLY ADAMS
**(February 1977–July 1978, NBC)**

James Adams (Dan Haggerty) is falsely accused of killing a man, and heads to the wilderness to hide. There he lives in a cabin, communes with Mother Nature, makes friends with a grizzly bear named Ben, and thus gets the nickname Grizzly.

The show was loosely based on the life of the real James "Grizzly" Adams, who was born in Massachusetts and spent a good deal of his adult life out West, in California's High Sierras, where he met the real Ben.

On the show, the setting is somewhere in the West, but *The Life and Times of Grizzly Adams* filmed extensively in and around Park City, Heber City, and the heavily timbered Uinta and Wasatch mountain areas of Utah.

The show also did location shoots in Arizona and New Mexico.

### WHAT TO SEE IN PARK CITY/HEBER AREA, UTAH

▶ **Mayflower,** off U.S. Highway 40 (about nine miles north of Heber). In this former silver mining area there are some old mining

The Mayflower Mine. *(Photo courtesy of the Mountainland Travel Region of Utah.)*

buildings that were used on the show. They are located on the banks of the Jordanelle Reservoir. The Mayflower Mine was part of the huge silver ore mining industry of the Park City area. Tourism is now the number one industry in the region, and plans are to develop Mayflower as a ski resort.

For tourism information in Heber, call the Heber Valley Chamber of Commerce, 801-654-3666.

For Park City tourism information, call the Park City Chamber Bureau, 800-453-1360.

### Fun Facts

• Denver Pyle, who played Mad Jack, was later Uncle Jesse on *The Dukes of Hazzard.*

• The show's characters were first introduced in a movie, also starring Dan Haggerty and Denver Pyle.

• Dan Haggerty worked as an animal trainer before becoming an actor, and was said to have a good relationship with his beastly costar Bozo, who played Ben the bear.

## THE LONE RANGER

**(September 1949–September 1957, ABC)**

"Hi-ho Silver, away!"

As every fan of this old western knows, Silver is the Lone Ranger's Horse, and also the color of the bullets he carries. The Lone Ranger (Clayton Moore and John Hart) never, however, shoots to kill.

He is really a Texas Ranger, John Reid, who with five fellow rangers was ambushed by desperadoes. The others died, but Reid recovered from his wounds with the help of his friend, Tonto (Jay Silverheels), his "faithful Indian companion." And he vowed to spend the rest of his life fighting evil.

The real location for some of the Ranger's good deeds was Utah (on the show it was the generic Old West). The show visited Kanab, Utah, several times in the 1950s, for location shooting.

Kanab, by that time, was already an established setting for western movie-making. The first movie shot in the town of 3,500, was in 1927, and Kanab has had an active film industry ever since, earning the nickname "Little Hollywood."

The area is ripe with attractive scenery, with all or part of four national parks, Zion, Bryce Canyon, Capitol Reef, and Canyonlands, located in the region.

Little Hollywood also offers tourists some kitschy attractions, reflecting its movie and TV heritage. Two large establishments compete for the tourist crowds with memorabilia collections, western-style meal offerings, and movie reenactments.

## What to See in Kanab, Utah

▶ **Kanab Canyon** (five miles north of Kanab on Highway 89). The canyon is located in a privately owned area with caves, Indian petroglyphs, and Indian ruins. The area is also home to Best Friends, one of the largest animal sanctuaries in the country. In addition to *The Lone Ranger,* shows that have gone on location here include *Death Valley Days*, *Wagon Train*, and *The Six Million Dollar Man*. Tours of the area are offered by jeep. Call 801-644-2001.

▶ **Kanab Movie Fort** (seven miles southwest of Kanab). The fort is on private property and is not accessible to the public. It is mentioned here for its historical significance, however. Built for movie location shoots, and now in ruins, the fort was used as a location for *The Lone Ranger,* as well as for *F Troop* (when the soldiers visited other forts). And it can be seen in at least one episode of *Lassie*. The road leading to the fort appeared on *The Adventures of Rin Tin Tin*.

▶ **Denny's Wigwam,** 78 East Center Street. This large Indian trading post and western-wear store has movie and television memorabilia including wagons and stagecoaches from *Gunsmoke* and *The Lone Ranger,* as well as a collection of movie guns and costumes. A western-style village out back is the setting for cookouts and movie reenactments, with audience members asked to take a role. The extravaganza (it's not just a store) is open daily, 9 A.M. to 9 P.M. Call 801-644-2452.

▶ **Lopeman's Frontier Movie Town and Western Heritage Museum,** 297 West Center Street. This museum/food hall has memorabilia from hundreds of movies that have filmed on location in the area including the set used on the movie *The Outlaw Josey Wales*, starring Clint Eastwood. A replica of an Indian village, built by local Navajos, is located out back, and there is also a collection of "old western stuff," as the proprietor described the collection. Both western-style meals and gunfight reenactments are offered for groups on a reservation-only basis (although individual travelers can sometimes join in too). The museum is open daily, and a small fee is charged for adults (kids are free). There is also a gift shop. Call 801-644-5337, or 800-551-1714.

▶ **The Parry Lodge,** 89 East Center Street. This eighty-nine-room lodge, which actually consists of several buildings, was built to house the actors and crews on the movie and TV projects in the area. The cast of *Gunsmoke* and *The Lone Ranger* are among those that stayed at the property, and there are autographed photos on the walls of the stars of these and other shows. The lodge has a special room dedicated to Ronald Reagan, who stayed at the hotel while he was on *Death Valley Days*. In the summer high season, rates range from $46 to $71 a night. Call 801-644-2601.

For area visitor information stop by the **Kane County Visitor Center,** 41 South, 100 East. Call 801-644-5033.

## Fun Facts

• Tonto's (Jay Silverheels) horse was named Scout. His "You, kemo sabe," was reserved for The Long Ranger. (It means trusted friend).

• That's the "William Tell" Overture played in the background in the show's intro, while announcer Fred Foy reads "A fiery horse with the speed of light . . ."

• Fred Foy later worked as an announcer on *The Dick Cavett Show.*

• The Lone Ranger was created originally for radio, in the early 1930s, by the same team that created The Green Hornet.

• Almost twenty years after the show went off the air, actor Moore went to court and, after a long legal battle, won the right to wear the famous Lone Ranger mask at public appearances.

## MORK & MINDY
**(September 1978–August 1982, ABC)**

There have been many sightings of UFOs in the Colorado Rockies, maybe because that cool mountain air can go straight to one's head. And due to the popularity of the sport of space-watching here, the producers of *Mork & Mindy* found the setting a natural for their Orkan-meets-Earth-woman sitcom. The fact that Boulder is a beautiful college town with lots of young people and historic architecture didn't hurt either.

The Victorian house seen on the show as Mindy's (Pam Dawber) apartment building, where she takes in Mork (Robin Williams) as a permanent guest, is really a single-family house in downtown Boulder. The real address of the house was mentioned on the first two episodes of the show, but the writers had to stop mentioning the address after droves of fans of the popular series descended on the real house, upsetting the family that lived there.

But the visitors kept coming. And in May 1979, a group of neighborhood children, led by Jonathan Ottman, whose parents owned the house, discouraged sightseers with homemade signs that said "Mork Doesn't Live Here So Go Away." Similar complaints were not heard at the real-life New York Delicatessen, which saw business surge after it was introduced on the show in the 1979 season.

While most of the production for the sitcom took place on a Hollywood set, the cast of *Mork & Mindy* visited Boulder for filming on several occasions, drawing crowds of fans. Denver was also used as a location. In one episode Williams joined the Pony Express cheerleaders for a taping at Mile High Stadium in Denver.

### What to See in Boulder, Colorado

- **McCallister House,** 1619 Pine Street. Exteriors only of this private Victorian house are seen on the show. The second-floor room, with the bay windows, Mindy's apartment on the show, is actually a bedroom. The house is an historical landmark. It was built by a lumberyard owner, Ira McCallister, in 1883. It's not far from the historic Hotel Boulderado, and just a few blocks from the Boulder Mall area.

*Mork & Mindy* lived here. *(Photo courtesy of the Boulder* Daily Camera.*)*

▶ **New York Delicatessen,** 1117 Pearl Street (on the Boulder Mall). The exterior of this restaurant appears as itself on the show. Inside there are a few pictures on the walls of the stars of the show. Depending on the season, the restaurant is frequented by locals, tourists, and/or college students. The menu is deli, including bagels and lox, corned beef, pastrami, and hamburgers. The deli is open daily, 8 A.M. until around 9 P.M., and later on busy nights. Call 303-449-5161.

▶ **The Downtown Boulder Mall on Pearl Street,** between 11th and 15th streets. This pedestrian shopping area, seen on the show, features galleries, boutiques, sidewalk cafes, and street performers.

▶ **Nature's Own Imagination,** 1133 Pearl Street. This is the building seen as McConnell's Music Store in the first season of the show. It's now a retail store with items made from natural materials including fossilized bones, amethyst, wood carvings, and gardening supplies. Global Response, an environmental group, is also headquartered here. Call 303-443-4430.

▶ **Boulder Bookstore,** 1107 Pearl Street. At the time of *Mork & Mindy,* this bookstore was the occupant of 1133 Pearl Street.

A crew from *Mork & Mindy* shoots in downtown Boulder. *(Photo courtesy of the Boulder* Daily Camera.*)*

When in town for filming, Williams bought books here, says owner David Bolduc. Call 303-447-2074.

▶ **Historic Boulder.** The preservation association offers walking tours of the downtown historic district. Call 303-444-5192.

## Fun Facts

• Mork (Robin Williams) and Mindy's (Pam Dawber) son was Mearth (Jonathan Winters), and he was half earthling and half Orkan. Mork gave birth to him in 1981. The name Mearth was a combination of the words Mork, Mindy, and Earth.

• In a 1981 episode called "Mork Meets Robin Williams," comedian Robin Williams starred as both himself and character Mork.

• This item appeared in the *New York Post* on December 20, 1979: "The Shah (of Iran) sent for about 20 t.v. [*sic*] cassettes of the hit *Mork & Mindy* series. He wants them to help him wile away the hours of his Panamanian exile."

• In Orkan "Na-nu! Na-nu!" means goodbye! "Shazbat!" means drats!

• Orkans drink with their fingers. They spit to say thank you, and they sit on their faces.

• Ork is about 200 million miles from Earth and has three moons.

• Actor Robin Williams studied acting at The Julliard School in New York.

• Orson (voice by Ralph James), Mork's boss on Ork, was heard but never seen on the show.

• The Orkans are enemies of the Necotons.

• Jay Thomas *(Love and War)*, left his job as a radio disc jockey in New York to appear on the show as Remo DaVinci, the manager of the New York Deli.

• Mork (Robin Williams) was first introduced on *Happy Days*. The Orkan landed in 1950s Milwaukee on the earlier show, and while Orkan Mork was not successful in his attempt to kidnap Richie (Ron Howard), Robin Williams was a hit, and the spin-off *Mork & Mindy* was created.

## FYI: ALSO IN THE WEST

Nearly three decades after the lawyer character of Perry Mason was first introduced on TV, Canadian-born actor Raymond Burr reprised his famous role regularly in TV movies under the title *The Perry Mason Mysteries* (December 1985 to November 1993, NBC).

The first of the movies, *Perry Mason Returns*, was one of the highest-rated TV movies of the 1985–1986 season, and twenty-five more were produced before Burr died from cancer at age seventy-six in September 1993 (he bravely worked despite his illness).

Perry Mason was based in L.A. in the 1950s and 1960s, but on the later show he did much of his deductive reasoning in Denver, Colorado.

### What to See in Denver, Colorado

▶ **The Equitable Building,** 730 17th Street (at Stout). This restored downtown historic building was seen in exterior shots as the location of Perry Mason's law offices. Inside is a marble lobby with a mosaic roof and stained-glass windows, and the building really does house the offices of lawyers, as well as other businesses.

▶ **The Denver Post Office and Federal Courthouse Building,** 18th Street and Stout. This is the real courthouse seen on the show. The large-columned government building, built in 1910, houses the U.S. Court of Appeals, Tenth Circuit.

▶ **The Denver City and County Building,** 1437 Bannock. Courtrooms 1 and 3 on the second floor of this courthouse building were both used by the show for interior location shoots. The rooms were later replicated on a set.

While the setting was supposed to be western Kentucky around the time of the Revolutionary War, *Daniel Boone* (September 1964 to August 1970), starring Fess Parker, used as a location Duck Creek Village, Utah.

The little town is up in the mountains, at about 8,300 feet, about a forty-minute drive from Kanab.

## What to See in Duck Creek Village, Utah

▶ **Duck Creek Village,** Highway 14 (from Kanab, drive north to Highway 89 and west on Highway 14). The population of this town is sixty. Besides being a location for the show, other attractions include cross-country skiing in the winter.

# ARIZONA AND NEVADA

Some great surprises await TV fans in this western region. You can actually visit the Ponderosa Ranch *(Bonanza)* in Incline Village, Nevada, and check out what's cooking in Hop Sing's kitchen; do some trading at Fort Courage *(F Troop)*, in tiny Houck, Arizona; and try your luck at the real casinos at the Desert Inn in Las Vegas *(Vega$)*.

Numerous TV western shows, from *Death Valley Days* to *The Young Riders*, have gone on location to Arizona, partly because it's just a short hop from L.A., and partly because the state's scenery is so impressive. A lot of the location shoots have taken place at Old Tucson, a movie location and western amusement park, just outside of Tucson, Arizona.

Old westerns, including *Death Valley Days*, also have gone on location to Apache Land, the old western movie town near Apache Junction, Arizona. While the Apache Land set is now closed, the nearby setting will bring back western TV memories.

*Alice*, the waitress sitcom, was based in Phoenix, but only a touch of the city was seen in its intro, including Scottsdale Road (which runs from Scottsdale to Carefree).

Similarly, *The New Dick Van Dyke Show*, which was shot in Carefree, Arizona, (where the star lived at the time), did not show much of the local scenery. It was done mostly inside a studio.

The scenic splendor of the Grand Canyon has been highlighted in several shows that have gone on location to the national park, including *Lassie* and *The Brady Bunch*.

*The Oregon Trail,* the short-lived western series, did location work in and around Flagstaff, Arizona, in the 1970s. And *Bonanza* filmed some forest scenes in the Coconino National Forest near Flagstaff, where the world's largest stand of ponderosa pines is located.

But *Bonanza,* of course, belongs mostly to Nevada.

So much was the demand by tourists for a real place to visit, production for the show actually moved from Hollywood to the Lake Tahoe, Nevada area. And the sets for the popular show are still an attraction today.

The short-lived *High Mountain Rangers,* with Robert Conrad, also filmed locations in Lake Tahoe.

Glitzy Las Vegas has not exactly been ignored by Hollywood, either. *Vega$* shot on location in the city, as did *Crime Story,* for a time, and numerous shows have visited the casino capital. *Designing Women* filmed episodes at the Tropicana Hotel, and this was where Anthony met his bride. *I Love Lucy* featured the Sands Hotel in an episode guest-starring Fred MacMurray and his wife, June Haver. *Hill Street Blues* went on location to the Desert Inn, *Remington Steele* to the Golden Nugget, and *Mike Hammer* to the Las Vegas Hilton.

Local sources said only once did a hotel in Las Vegas not appear as itself, and that was when the Las Vegas Hilton changed its name to go with the plot of a Jackie Collins miniseries.

In little Houck, Arizona, near the New Mexico border, the set for *F Troop* still stands, and it's open to visitors who stop by the adjacent gift shop.

Other regional attractions of interest to TV trivia buffs include The Rex Allen Museum, honoring Rex Allen, star of the 1950s show *Frontier Doctor,* in Willcox, Arizona.

And in Kingman, Arizona, there's an annual tribute in October to hometown celebrity Andy Devine of *Andy's Gang.* There's also a permanent display with Devine memorabilia at The Mohave Museum of History and Arts. Devine also appeared for a year on *Flipper.*

## BONANZA

**(September 1959–January 1973, NBC)**

The original Ponderosa Ranch was a Hollywood set. But the burning map at the beginning of each episode of *Bonanza* pinpointed the location of the fictional ranch as being on the north shore of Lake Tahoe, in Nevada, and the western proved so popular that fans refused to believe there wasn't a real Ponderosa Ranch. They came to Lake Tahoe looking for the location.

The burning map originated with a Hollywood Studio artist who was told by show creator David Dortort to come up with something that would "burn Bonanza into the minds of the viewers . . ."

Obviously it worked.

Seeing an opportunity in 1967, Bill Anderson, a California equipment contractor, and his wife, Joyce, decided to create a real Ponderosa Ranch. Actors Lorne Greene (Ben Cartwright), Dan Blocker (Hoss), and Michael Landon (Little Joe) all signed up as partners in the venture. The setting at Incline Village, Nevada, at the exact location of the spot on the burning map, became the location for filming of the show (for six seasons), as well as a popular tourist attraction.

*Bonanza* also used other location sites for filming, right from the beginning, shooting on location in California, Nevada, and Arizona. And the use of colorful locations was not accidental. The show was the first network western filmed in color, and was developed explicitly to encourage the sale of color television sets. RCA was the parent company of NBC, and wanted people to buy its new products.

Other locations seen on *Bonanza* include Old Tucson, near Tucson, Arizona, the Coconino National Forest in Flagstaff, Arizona, and various ranch sites outside of L.A., including the now defunct Iverson's Movie Ranch in the San Fernando Valley.

### What to See in Incline Village, Nevada

▶ **The Ponderosa Ranch,** 100 Ponderosa Ranch Road. The setting is pretty much the way it was during filming of the show (in fact the location was also used for later television movies based on

The Ponderosa Ranch set. *(Photo courtesy of The Ponderosa Ranch.)*

the show). There is the Cartwright Ranch House and Hop Sing's kitchen, complete with furnishings from the show, including the hats and coats of Little Joe and Hoss. Because this is a tourist attraction there is also a theme park with shops, rides, a fun arcade, and saloons. You can eat Hossburgers and drink from a souvenir tin cup (see below), or have your photo taken in western gear. Also offered are guided horseback tours that show locations used for filming. For younger visitors there is a Pettin' Farm and Hoss Bear (a costumed character). There's a wedding chapel at the Church of the Ponderosa, for those inclined to western-style nuptials. The ranch is open May to October (although there has been some talk of going year-round), from 9:30 A.M. to 5 P.M. Call 702-831-0691.

### Fun Feature: The Old Tin Cup

Most visitors to the ranch leave with a tin cup souvenir. The idea came from ranch owner Bill Anderson, who got the idea from actor Lorne Greene's popular recording of "An Old Tin Cup," which was on the charts around the time the ranch was opened.

Everything you buy to drink at the ranch comes in the cups, including soft drinks, snow cones, and the ranch's own brand of Kentucky bourbon.

"The tin cup is by no means an art object," Anderson said, "but people seem to want to identify with cowboys, the Old West, and something out of the past."

The wide-scale use of the cups at the Ponderosa Ranch has actually made the ranch the largest tin cup customer since the Civil War. The cups are produced by Schlueter Manufacturing Company of St. Louis.

## Fun Facts

- About a quarter of the scripts for *Bonanza* were written by amateurs, including a former Berkeley, California policewoman, who wrote the episode called "Erin"; a Van Nuys, California, housewife; a student who had studied scriptwriting with Rod Serling *(The Twilight Zone)* at Ithaca College in New York; and a stuntman who had worked on *Bonanza.*
- *Bonanza* was broadcast in eighty-six foreign countries in twelve languages. It was believed to be shown in bootleg fashion in the U S.S.R. In Japan, the sound track was dubbed in classical Kabuki theater style.
- Dan Blocker, who played Hoss, had turned down an offer to play pro football to be an actor, and had earlier appeared as a heavy on *Gunsmoke.*
- The show was so popular that President Johnson would not schedule presidential speeches on Sunday at nine P.M., because he reportedly thought too many viewers might be offended. Dan Blocker—like Johnson, a Texan—once visited LBJ at the White House. The Queen of England was believed also to be a fan of the show.
- Actor Lorne Greene, who played Ben, had previously appeared in an episode of *Wagon Train.* But he was known to Canadians as the Voice of Canada, for his broadcasting work during World War II.
- Michael Landon, who played Little Joe, was born on Long

The Cartwright's place. *(Photo courtesy of The Ponderosa Ranch.)*

Island and was also an athlete. He set a national high school record in the javelin, but was sidelined at USC when he hurt his arm.

- Pernell Roberts, who played Adam, left the show in 1965 because he said it compromised his artistic freedom. He told *The New York Times*, "My being part of *Bonanza* was like Isaac Stern sitting in with Lawrence Welk." When Roberts left the show, the character of Adam left to study abroad.
- Dan Blocker also thought about leaving in 1965, telling a Hollywood trade paper that he was just playing Hoss for the money and felt like "a true prostitute." Blocker stayed with the show, but died from a blood clot in the lung before the 1972 season.
- Hoss was originally scheduled to get married on the show, but when the actor died the script was rewritten so that Little Joe (Landon) was the bridegroom. Bonnie Bedelia played the

bride in the wedding episode, which was written by Landon, and filmed on location in Sonora, California.

• Andy Robinson, who appeared in the two-hour wedding episode as Joe's brother-in-law, had previously gained recognition as the psychopath in the film *Dirty Harry*.

• A guest on the show in 1968 was Julie Harris. A big star at the time, Harris said she took the role because she was such a big *Bonanza* fan. She played a married woman who falls for Hoss.

• Executive producer David Dortort said he modeled Ben after his own father, whose name was also Ben.

• David Canary, who later was Adam and Stewart Chandler on *All My Children*, joined the cast of *Bonanza* in 1967, as Candy, a ranch hand. A one-time draft choice of the Denver Broncos, Canary also opted for an acting career. He was hired in part for his nose, which he had broken playing football.

• David Cassidy appeared on *Bonanza* right before *The Partridge Family* made him a teen idol. He played a young man accused of murder in an episode called "The Law and Billy Burgess."

• In 1972, *Bonanza* was honored by the Western Apparel and Equipment Manufacturers Association in Denver for keeping the public western-oriented.

• In 1967 and 1968, a portable duplicate of the Ponderosa, called The Bonanza Ranch House Caravan, toured state fairs, including those in Arizona and Springfield, Illinois. Lorne Greene was behind the creation of the exhibit.

• Besides becoming millionaires, the stars of the show each got to pick a new model Chevrolet, courtesy of sponsor, General Motors, each year.

## Fun Feature

There was no real Cartwright family, but it was hard to tell people in Nevada that. During filming of *Bonanza* the stars and producers were asked numerous times about the Cartwright family.

Executive producer David Dortort said one old-timer approached him in a restaurant and asked if the actors in the show were related to the "old Cartwright" family.

"I told the old fellow that there never was a real Cartwright family—and he got so mad. I thought he was going to hit me," Dortort said.

Lorne Greene once gave his autograph to an elderly woman who insisted she knew real Cartwrights, and who studied Greene's signature as tears welled in her eyes. She told the actor, "You not only look just like your pa did—but you even write like he did."

Dan Blocker said someone once asked him for a loan, claiming to have earlier worked on the Ponderosa Ranch.

## F TROOP
**(September 1965–August 1967, ABC)**

The happy band of misfits in this western sitcom were based at Fort Courage, somewhere in the West, after the Civil War. Captain Wilton Parmenter (Ken Berry), the bumbling commanding officer at the fort, is a hero of the war, having accidently led a charge against the enemy (he was trying to leave but went the wrong way).

Really controlling happenings at Fort Courage, however, is Sergeant O'Rourke (Forrest Tucker) and his sidekick Corporal Agarn (Larry Storch). The two have a pact with the nearby Hekawi Indian tribe that guarantees peace as long as the cavalrymen can keep orders for Indian souvenirs coming in. Everyone wants to make a buck. The Indians on the show are friendly enough to stage uprisings only when asked, and usually to impress some visiting representative of Uncle Sam.

Two sets were used as the fictional Fort Courage, one in Hollywood and the other in a tiny town in northeastern Arizona.

Given the plot of the show, it's fitting that today the fort is attached to a trading post that specializes in Indian jewelry and other Native American handicrafts.

### What to See in Houck, Arizona

▶ **Fort Courage,** I-40 (about thirty miles west of Gallup, New Mexico, between Lupton and Chambers, Arizona). The fort set is owned by Armand Ortega, a New Mexico businessman, who operates a trading post, restaurant, and service station at the site. Visitors can climb the look-out tower, seen so often on the show. Souvenirs for sale in the store include *F Troop* items. Ortega also owns the famous El Rancho hotel, which has housed many movie stars, in nearby Gallup, New Mexico, since 1937.

### Fun Facts

• Guest stars on *F Troop* included Paul Lynde, who appeared as a singing Royal Mountie, and Don Rickles and Milton Berle, who played Indians, named Bald Eagle and Wise Owl, respectively.

• Larry Storch not only played Corporal Agarn on the show, but also his look-alike cousin, Lucky Pierre.

• Wrangler Jane (Melody Patterson) provided the love interest for Captain Parmenter (Ken Berry), and there were reports that when the show started, Patterson was only sixteen. Other reports had her at eighteen, however.

• Actor Ken Berry later appeared on *Mayberry R.F.D.* and *Mama's Family.*

## LASSIE
**(September 1954–September 1971, CBS, then syndicated until 1974)**

Lassie was always loyal to her masters, whoever they were. The noble collie fought wrongdoing and protected humans from potential tragedy on TV, on Sunday nights, for seventeen years.

The famous dog was first introduced in a best-selling book, *Lassie Come Home*, and appeared in movies and on radio before her TV stardom.

On TV, Lassie first lived with the Millers, including young Jeff Miller (Tommy Rettig) on a farm near Calverton. In 1957, the Millers moved to the city, selling the farm to the Martins. The Martins adopted a young orphan named Timmy (Jon Provost), who became Lassie's top companion.

In 1964, the Martins decided to move to Australia, with Timmy, but without Lassie. Timmy's friend, elderly Cully

(Andy Clyde) watched the dog for a time and then turned her over to U.S. Forest Service Ranger Corey Stewart (Robert Bray). After Ranger Stewart was injured on the show, Lassie became the companion of fellow Rangers Scott Turner (Jed Allan) and Bob Erickson (Jack de Mave).

Later Lassie was on her own for a time, and she met a mate and had puppies in 1970. But in 1972, she ended up on the California ranch of the Holden family.

*Lassie* was initially filmed mostly in a studio in L.A., although the series pilot was shot in Calgary, Canada. Later, location shooting increased; the 1961 season premiere episode was filmed at the Grand Canyon.

According to *Lassie: A Dog's Life*, a book by Ace Collins, the Grand Canyon was also used for other winter scenes. Since the show filmed mostly in southern California, the cast and crew went to the Grand Canyon, and other locations, to find snow. When this was not possible, fake snow, made of soap, was sometimes used on the set, but it was confusing for the Lassie dog, who would try to eat the mixture, Collins writes.

Winter at the Grand Canyon. *(Grand Canyon National Park, photo no. 3171.)*

Railtown, a California state historic park in Jamestown, California, also appeared several times on the show. And in 1963, the show filmed a five-parter in the High Sierras (which was later combined to make the movie *Lassie's Great Adventure*).

When Lassie started working with the forest rangers in 1964, she increasingly traveled on location to national parks and other settings around the country.

## What to See at the Grand Canyon, Arizona

▶ **Grand Canyon National Park** (on Route 64). Among shoots here was "Lassie at the Grand Canyon," an episode that aired in 1961. According to park records at the time, the script had to be reviewed and revised by the Grand Canyon National Park staff to make sure it adhered to park policies and regulations. It's believed the South Rim, which is open year-round, was used for the location shoot. The South Rim averages 7,000 feet above sea level. For general park information, call 602-638-7888.

## Fun Facts

• Lassie's trainer, Rudd Weatherwax, got the first Lassie collie puppy for $10.

• The original Martins were played by Cloris Leachman *(The Mary Tyler Moore Show)* and Jon Shepodd. But Timmy got new actors as parents in 1958, June Lockhart and Hugh Reilly.

• Lassie had a movie career before her TV debut, including *Lassie Come Home*, which featured a young Elizabeth Taylor and Roddy McDowall. Taylor also appeared in the third Lassie movie, *Courage of Lassie*. Actor Peter Lawford made his film debut in *Son of Lassie*.

• Tommy Rettig, who played Jeff, grew up to become a creator of computer software programs, according to writer Ace Collins in *Lassie: A Dog's Life*. Jon Provost, who played Timmy, went into real estate and did volunteer work for Canine Companions for Independence.

• Lassie was supposed to be a female dog, but there were six dogs seen as Lassie on TV, and they were all males. Male dogs were used because females tend to lose their coats when they are in heat, and also because male dogs are bigger, according to writer Collins.

• While on TV, Lassie made lots of personal appearances to meet her fans, traveling to county fairs, amusement parks, theaters, and parades. The collie even performed at Madison Square Garden.

• When Lassie traveled, she flew by plane if the trip was more than 300 miles, always in first class. Her trainer, Rudd Weatherwax, would buy two seats. Weatherwax would take one, and Lassie would stretch out on the floor underneath the other.

• A cartoon version of Lassie, *Lassie's Rescue Rangers*, aired from 1973 to 1975. *The New Lassie* was produced for syndication from 1989 to 1991.

• Larry Wilcox, who later starred in *CHiPs*, appeared in *Lassie* from 1972 to 1974, as Dale Mitchell, Lassie's last owner.

• The show was also called *Jeff's Collie.*

• Campbell's Soup sponsored the show, and kitchen shots on *Lassie* showed plenty of soup cans on the shelves.

• Lassie merchandise has become popular with collectors; hot items include books, posters, toys, and *TV Guide* issues with Lassie's photo, according to *Antiques & Collecting* magazine.

## LITTLE HOUSE ON THE PRAIRIE
**(September 1974–March 1983)**

This family drama, the baby of Michael Landon *(Bonanza)*, who was executive producer, star, and sometimes writer and director of the show as well, managed to pull at the heartstrings in nearly every episode. Landon played Charles Ingalls, a farmer who moved his family of five from the plains of Kansas to the untapped western town of Walnut Grove, Minnesota, in the 1870s.

The show was based on the writings of Laura Ingalls Wilder, who is credited with writing nine books, the first published in 1932, about her own family's experiences settling in the West.

On the show, actress Melissa Gilbert played daughter Laura, who was also the show's narrator. Landon sometimes enhanced the stories of the Wilder books with events based in his real

On the set of *Little House on the Prairie.* *(Photo courtesy of Old Tucson Studios, Tucson, Arizona.)*

life. For instance, as religion grew important in his own life, Landon introduced prayer on *Little House on the Prairie.*

Roger MacBride, a Virginia lawyer who inherited the rights to *The Little House* books from his adopted grandmother, Rose Wilder, daughter of Laura Ingalls Wilder, picked Landon to do the TV adaptation. But MacBride later complained that the show did not always realistically depict the life of the Ingalls family. He noted in interviews, for instance, that on the show, Charles Ingalls (Landon) served his kids orange juice for breakfast in the winter, and that they wore shoes to school everyday, which would not have been likely in the real-life 1870s.

On the show, in 1978, the Ingalls family moved away from the farm, settling briefly in the city of Winoka, before returning to Walnut Grove.

*Little House on the Prairie* did not film in Minnesota, instead going on location several times, between 1977 and 1983, to the western movie sets of Old Tucson and Mescal, near Tucson, Arizona. Both Michael Landon and producer Kent McCray had previously worked in Old Tucson, known as the Hollywood of Arizona, on *Bonanza.*

The show shot mostly in California, however, in a studio, and at a movie ranch in Simi Valley (which is not open to the public). Occasionally other California locations were used as well, including Jamestown (for train scenes) and Sonora, California (for river and lake scenes), according to Producer McCray.

Much of the show's set was destroyed by dynamite in a farewell episode, in which the residents of Walnut Grove destroy the town rather than turn it over to a robber baron.

## What to See in Tucson, Arizona

▶ **Old Tucson,** 201 S. Kinney Road (about fourteen miles west of downtown Tucson, off Interstate 10). Okay, it's a bit touristy. But Old Tucson is a must for western fans. And there are a lot of you out there—this western movie set and family fun park gets about a half million visitors a year, making it Arizona's second most popular tourist attraction after the Grand Canyon. The first movie filmed here was *Arizona* in 1939, and the 365-acre site, which neighbors the Saguaro National Monument, averages twenty-five movie and TV productions a year (many great westerns have filmed here). Visitors can watch productions in progress, and sometimes catch a glimpse of the stars. The park also features games and rides as well as stunt demonstrations, staged gunfights, and musical reviews. And there are museums and exhibits, as well as restaurants and gift shops selling western items. Old Tucson also runs Mescal, another western movie set that is not open to the public. Old Tucson is open 9 A.M. to 9 P.M. daily, year-round (except Thanksgiving and Christmas). Admission is charged. Reduced admission is offered at night. Call 602-883-0100.

A fire destroyed about forty percent of Old Tucson on April 24, 1995. The park is scheduled to reopen in 1996.

## Fun Facts

• Michael Landon's daughter, Leslie, joined the cast in its ninth season as the new school teacher, Etta Plum. Leslie also played a pregnant sixteen-year-old wife in a 1979 episode.

• The son of actor Dan Blocker, who appeared with Landon on *Bonanza*, and other Landon kids also appeared on *Little House on the Prairie*. Landon had nine children from several marriages.

• In addition to *Bonanza,* Michael Landon's acting credits include starring in the 1957 film *I Was a Teenage Werewolf.* He also filled in for Johnny Carson for a time on *The Tonight Show.*

• Michael Landon told interviewers his parents had had a "mixed marriage" of both height and religion, and in an autobiographical nod, *Little House on the Prairie* once featured as characters a short Jewish man married to a tall Christian woman.

• Jonathan Gilbert, the real brother of Melissa Gilbert, who played Laura, played Willie Oleson, the storekeeper's son.

• Little sister Sara Gilbert is also an actress. She played Darlene on *Roseanne.*

• Melissa Gilbert grew up to star in *Sweet Justice,* and to gain a reputation for having a broken heart. She had an on-again, off-again relationship with movie actor Rob Lowe and was divorced from actor/playwright Bo Brinkman. In high school, she dated Michael Landon, Jr. But she told *TV Guide* that "the most romantic night" of her life was January 1, 1995, when she married Bruce Boxleitner *(Scarecrow and Mrs. King).*

• When Michael Landon left the show in 1982, it was renamed *Little House: A New Beginning* and focused on the lives of Laura (Melissa Gilbert) and her husband, Almanzo (Dean Butler).

• Actress Shannen Doherty, who became a star on *Beverly Hills 90210* and earned a Hollywood-style "bad girl" reputation, played Jenny, Almanzo's young niece on later episodes of *Little House: A New Beginning.*

## FYI: OTHER MICHAEL LANDON SHOWS SHOT AT OLD TUCSON

*Father Murphy* (November 1981 to December 1982, NBC), a family drama created and produced by Michael Landon, and starring former football player Merlin Olsen, filmed in Old Tucson and the surrounding area. The schoolhouse featured on the show was destroyed in the park's 1995 fire. And the street seen as the main street in Jackson, the show's mining town setting, is really Kansas Street in Old Tucson.

On the set of *Father Murphy*. *(Photo courtesy of Old Tucson Studios, Tucson, Arizona.)*

The pilot for Michael Landon's *Highway to Heaven* (September 1984 to June 1988, NBC) was also filmed at various sites in Old Tucson and the surrounding area. Most of the show's other episodes were also shot on location in sites including the El Toro (California) Marine Base and Camp Good Times in Malibu, California, a camp for children with cancer.

## FYI: OTHER SHOWS SHOT AT OLD TUCSON

The western *The High Chaparral* (September 1967 to September 1971, NBC) was filmed on sets in both Old Tucson and Hollywood. The High Chaparral Ranch House & Barn is part of the Old Tucson studio tour, as is the Red Dog Hotel, which appeared as a saloon on the show.

*The Young Riders* (September 1989 to July 1992, ABC), which focused on the lives of six young Pony Express riders, filmed in both Old Tucson and Mescal, an area operated by the Old Tucson studio, but closed to the public.

*The High Chaparral* set at Old Tucson Studios. *(Photo courtesy of Old Tucson Studios, Tucson, Arizona.)*

On the set of *The Young Riders*. *(Photo courtesy of Old Tucson Studios, Tucson, Arizona.)*

At Old Tucson, the show used permanent sets including the Mexican Plaza & Mission building (which also appeared on an episode of *Hart to Hart*) and the Lincoln County Court House. Both the Mission building and the courthouse were destroyed in the park's 1995 fire. The fort wall at Old Tucson's Fort Reunion was built specifically for *The Young Riders*, in 1989.

## VEGA$
**(September 1978–September 1981, ABC)**

The main character of *Vega$*, Dan Tanna (Robert Urich), lived and worked at the Desert Inn in Las Vegas. And lots of location footage of the real hotel and the glitzy city appeared on the show.

The real Desert Inn, now called the Sheraton Desert Inn Resort & Casino, was seen in interior and exterior shots, including the Wimbleton Penthouse, which was, on the show, the office of Philip Roth (Tony Curtis), Tanna's millionaire boss. Also featured were the hotel's casino, lounge area, lobby, pool area, health spa, and tennis courts.

Tanna also visited other famous Vegas hotels including Circus Circus Hotel & Casino, The Golden Nugget Hotel, and Caesars Palace.

The neon-lighted Las Vegas Strip is prominently shown on *Vega$*, as are many other sights in the city including the downtown area. Car chases were filmed on the Strip and on side streets. Interiors were also done on a set in L.A., and the show occasionally visited other locations for shoots, including Hawaii. Real staff members of the Desert Inn hotel appeared in scenes on *Vega$*, as did some guests.

### What to See in Las Vegas, Nevada

▶ **The Sheraton Desert Inn Resort & Casino,** 3145 Las Vegas Boulevard. The Wimbleton Penthouse Suite, Philip Roth's place on the show, is 4,000 square feet and priced at more than $1,500 a night. But the suite, which is located in a seven-story pyramid structure between the hotel's main building and the golf course, is not available to the general public. Guests have 820 other rooms to choose from, however, with rates starting at $125 per

The Sheraton Desert Inn Resort & Casino. *(Photo courtesy of the Sheraton Desert Inn Resort & Casino.)*

The casino at the Sheraton Desert Inn Resort & Casino. *(Photo courtesy of the Sheraton Desert Inn Resort & Casino.)*

night. Anyone wanting to make a wager is invited to visit the hotel's twenty-four-hour casino, of course. The hotel also has five restaurants. Big stars of the likes of Liza, Frank, and Dean perform in the hotel's theaters. Call 800-634-6906.

For Las Vegas tourism information, call 800-332-5333.

### Fun Facts

• The part of Binzer (Bart Braverman) was created for the pilot episode only, but the producers liked the chemistry between Braverman and Robert Urich, who played Tanna, and enhanced the part. Binzer's name on the show is really Bobby Borso.

• Guest stars on the show included Lorne Green and Pernell Roberts, who were both stars of *Bonanza*, which filmed for several years on location in Incline Village, near Lake Tahoe, Nevada.

• Detective Tanna (Urich) drove a red T-bird (circa 1959) convertible.

• Robert Urich was recommended for the role of Tanna by Burt Reynolds. Before becoming an actor, Urich was an executive at WGN Radio in Chicago. He also played Peter, the sexy tennis pro, on *Soap* and later starred in *Spenser: For Hire*.

• *Vega$* was coproduced by Aaron Spelling, who also coproduced *Charlie's Angels*, which *Vega$* followed in the ABC Wednesday night lineup for two seasons.

## FYI: ALSO IN NEVADA

On *Crime Story* (September 1986 to March 1987, and June 1987 to May 1988, NBC), mobster Ray Luca (Anthony Denison) moved from Chicago to Las Vegas as he saw his star rising. And tough cop Mike Torello (Dennis Farina) followed, still trying to catch his man.

Sets for *Crime Story* were created on a large sound stage in Las Vegas, and extensive shooting was also done at the historic Moulin Rouge hotel and at the massive Vegas World Hotel & Casino.

Other exterior shots of the city of Las Vegas, the Strip, and the surrounding landscape also appeared on the show.

## What to See in Las Vegas, Nevada

▶ **Moulin Rouge,** 900 W. Bonanza Road. Filming for the show took place inside and out of this older property, with the main showroom, small bar, restaurant area, and parking lot among the locations used. The hotel itself is noteworthy as the first interracial hotel in Las Vegas. Blacks performing on the strip were barred from the big casino hotels and instead stayed here. The hotel was built in 1955, and was recently declared an historic landmark for its role in the breaking of color barriers. The property has 116 guest rooms and eight apartments for both transient guests and permanent tenants. Live poker and slot machines are offered in the casino. Rates are $28 to $32 a night. Call 702-648-5054.

▶ **Vegas World Hotel & Casino,** 2000 Las Vegas Boulevard. This 1,000-room property was also seen on the show. Rates are $35 to $85 a night. Call 702-382-2000.

(See also Chicago).

# LOS ANGELES

It may seem like Los Angeles is the center of the universe for TV production. And it's true that most TV shows film interiors in the Hollywood area.

It's also true that if the plot of a show is based in L.A., it's probably L.A. buildings you're seeing in the background. Why should production crews leave the City of Angels?

L.A. has also doubled for other locales on TV shows (although not all, as the rest of this book shows).

A private house in Hollywood, for instance, was seen as the Cunningham family's Milwaukee, Wisconsin home on *Happy Days*, and the closed-down Linda Vista Hospital in East L.A. doubles for the Chicago hospital setting of *ER* (or did, before the show moved to its more permanent L.A. studio set).

And while it may seem like you're seeing Manhattan, the apartment building known as the home of neurotic New Yorker Jerry on *Seinfeld* is really also in L.A. (reportedly near Sixth Street and Vermont Avenue).

L.A. landmarks have doubled for fictional locations too. The real Los Angeles City Hall was seen as the *Daily Planet* building in Metropolis, on *The Adventures of Superman*. And while *Batman* takes place in Gotham City, the Batcave is really in L.A.'s Griffith Park.

Fiction also merges with reality on several shows in which L.A. buildings are in real life as they appear on TV, such as the office building seen on *L.A. Law* (where real lawyers work) and the police headquarters building seen on *Dragnet* (where real L.A. cops can be found).

There also really was a Dino's Lodge restaurant, as on *77 Sunset Strip*, but it's long gone (77 is a fictional address). And the parking lot where Kookie parked cars (really at 8433 Sunset Boulevard) also is no longer there.

*The Brady Bunch* fans will delight, however, in the fact that there really is a private house, in a quiet neighborhood of L.A., that was seen on the show as the home of Greg, Marcia, Jan, Peter, Bobby, Cindy, et al.

There is also a real Melrose Place in L.A., although not exactly as seen on *Melrose Place*. And the kids on *Beverly Hills 90210* are seen attending a real L.A. college, although it is not called California University.

Of course, the Hollywood area studios are the home base for most TV shows. Universal Studios (100 Universal City Plaza) is open as a tourist attraction, and offers a tram ride of production sites that sometimes includes TV show sets. Depending on what areas are being used for current productions (as a working studio, sets are often moved around), visitors might view the house the Cleaver family called home on *Leave It to Beaver*, the pond that doubled for the South Pacific on *McHale's Navy*, and the ghoulish home seen as that of Herman and Lily on *The Munsters*.

## THE ADVENTURES OF SUPERMAN
**(1951–1957, syndicated)**

Jerry Siegel and Joe Schuster, two teenagers from Ohio, created the character of Superman in the 1930s. And "The Man of Steel" appeared in Action Comics comic books, on radio, and in movies before making his TV debut. Film actor George Reeves played the TV Superman, as well as his alter ego, the "mild-mannered reporter," Clark Kent, who worked at the *Daily Planet* in Metropolis. The building seen as the *Daily Planet* on the show really exists in L.A., and is, in fact, Los Angeles City Hall.

In 1959, two years after the show went off the air, Reeves (whose real name was George Besselo) died in his Beverly Hills home, with the death ruled a suicide. There were some who questioned the ruling, denying Reeves feared, as rumor had it, that he had become too identifiable to the public as Superman and consequently would not be considered for other acting roles. Further investigations failed to prove another cause for the actor's death.

## What to See in Los Angeles, California

▶ **Los Angeles City Hall,** 200 N. Spring Street. The historic city hall building doubled as the *Daily Planet* on the show. It also appeared on *Dragnet,* as itself, and on numerous other television shows and movies.

Los Angeles City Hall. *(Photo by Fran Golden.)*

## Fun Facts

• Superman was "faster than a speeding bullet. More powerful than a locomotive. And able to leap tall buildings in a single bound."

• Superman's real name was Kal-El, and he was from Krypton, but was raised by humans in Smallville. He used his special powers to help mankind, fighting "a never-ending battle for truth, justice, and the American Way."

• Kryptonite was the only thing that could weaken Superman.

• The show was shot on a tight budget, and the actors were seen wearing the same costumes for several episodes. Only a few scenes of Superman flying were shot, and they were repeatedly shown on the show.

• Actor George Reeves earlier appeared in *Gone With The Wind.* His numerous film appearances also included *Knute Rockne—All American* (with Ronald Reagan) and *From Here to Eternity.*

• Actor Reeves was not as muscular as Superman, and he was seen in a suit padded with foam rubber when he appeared as the Man of Steel.

• Kellogg's was the main sponsor of the series.

• On *Lois & Clark*, a 1990s version of the Superman story, Phyllis Coates, who was one of two actresses to play Lois (Noel Neill was the other) on the original show, made a guest appearance as the mother of modern-day Lois (Teri Hatcher).

• Phyllis Coates told *TV Guide* in 1994 that she still gets mail from fans who believe there was a flirtation going on between her, as Lois, and George Reeves, as Superman. But she said, "They're projecting—believe me, we weren't playing it that way."

• Clark Kent's friends actually discovered his true identity in one episode, but they quickly forgot, with the help of an antimemory spray.

• The cartoon show *Underdog* played on the *Superman* intro, "It's a Bird! It's a Plane . . ." Instead of announcing "It's Superman!" Underdog (the voice of actor Wally Cox) said, "It's just little old me, Underdog!"

## BATMAN
**(January 1966–March 1968, ABC)**

The character of Batman, a superhero who is technologically powerful but has no supernatural powers, was created in 1939 by Bob Kane for Detective Comics. Batman was featured on the *Superman* radio show and in two movies before appearing on TV.

William Dozier, the producer of TV's *Batman*, and also the narrator of the show, said the show was designed for kids. But the purposely campy, comic-book-like presentation of the show also made it popular with adults. "Bam!"

Batman (Adam West) did his fighting of evil in Gotham City, with his sidekick Robin (Burt Ward), the Boy Wonder, and together the two were the Dynamic Duo. "Pow!" The show was mostly done in a studio, but included glimpses of a real-life Pasadena mansion, seen as the home of Batman's alter ego, Bruce Wayne. A real cave in Griffith Park doubled as the Batcave on the show. "Curses!"

The caves at Griffith Park. *(Photo by Ben Pivar and Jennifer Mencken.)*

Producer Dozier sought actors with classical training in drama to be the bad guys, and playing a villain on *Batman* became a status symbol in Hollywood. "Holy Barracuda, Batman!"

West took his role as the Caped Crusader very seriously. "I am deeply and humbly aware of the moral obligations that accompany my work in this series," he said. "WOW!"

## What to See in Los Angeles, California

▶ **Bronson Canyon Caves,** in Griffith Park (at the end of Canyon Drive). This is the Batcave, as seen on *Batman*. But in real life the area is marked "Rock Quarry" on local maps. Getting to the real caves requires a hike on foot, on a dirt road, for about ten minutes. The location also appears on *Gunsmoke, Bonanza, The Twilight Zone,* and *Star Trek: The Next Generation.* The park is open daily until 10 P.M. Call the ranger office at Griffith Park, 213-665-5188.

## What to See in Pasadena, California

▶ **Batman Mansion,** 380 S. San Rafael Avenue. This privately owned mansion appeared as Bruce Wayne's home on the show. It's in a quiet residential area and is not very visible from the street.

## Fun Facts

• Batman drove a Batmobile, often to the Batcave. He contacted Gotham City's Police Commissioner Gordon (Neil Hamilton) on the Batphone, and responded to the Batsignal (a searchlight with Batman's logo, used for emergencies).

• *Pravda,* the Soviet newspaper, called Batman "the representative of the broad mass of American billionaires."

• The show was initially broadcast twice a week (later once a week).

• Batgirl (Yvonne Craig), who joined the show in its second year, was also Barbara Gordon, the librarian daughter of the police commissioner.

• Actors Adam West and Burt Ward also did the voices of Batman and Robin for a 1970s cartoon version of the show.

• Well-known celebrities who appeared on *Batman* included Burgess Meredith as The Penguin, Otto Preminger and Eli Wallach as Mr. Freeze, Eartha Kitt and Lee Ann Merriwether as Catwoman, Vincent Price as Egghead, Frank Gorshin and John Astin as The Riddler, Cesar Romero as The Joker, and Tallulah Bankhead as the Black Widow.

• Former presidential press secretary Pierre Salinger got into the act playing Lucky Pierre.

• Adam West earlier appeared on *The Detectives*.

• The show inspired the sale of millions of dollars in merchandise, including *Batman* lunch boxes, dolls, and pajamas. And that was before the 1980s movie version!

## THE BRADY BUNCH
**(September 1969–August 1974, ABC)**

Created by Sherwood Schwartz, the man also responsible for *Gilligan's Island*, this happy family sitcom was not a big hit during its initial run. It never even made the annual top ten list of shows. But *The Brady Bunch* became a cult classic in reruns.

It was the first show featuring a merged family. Widower Mike, (Robert Reed) married smiley widow Carol (Florence Henderson), and she brought with her three daughters, Marcia, Jan, and Cindy, while he brought with him three sons, Greg, Peter, and Bobby.

The boys were brunettes like Dad (whose hair style seems to constantly change) and the girls were blond like Mom, except for one time when Jan (Eve Plumb) donned a brunette wig.

It's been estimated that more babyboomers know the words to the show's theme song than to the national anthem ("There's a story, of a lovely lady . . ."). So popular was the show that the Bradys have been parodied on *Saturday Night Live* and by various comedians, including Roseanne Barr, and have been mentioned in speeches, including those of President George Bush, when he was in office.

On the show, the Brady family lived in the L.A. area, in a four-bedroom house at 4222 Clinton Avenue. The real house seen on the show is a private home located near Universal City. Interiors were shot in a studio.

*The Brady Bunch* house. *(Photo by Fran Golden.)*

The show was on the air during the Vietnam War, but weighty issues like war were not mentioned. Politics were, however, but only on the level of Marcia (Maureen McCormick) running for student council president.

More important issues were how Greg (Barry Williams) was going to hide a goat in the attic, whether Cindy (Susan Olsen) would get over thinking she was Shirley Temple, or whether Bobby (Michael Lookinland) would get mumps after he was kissed for the first time.

Guest stars on *The Brady Bunch* included Davy Jones *(The Monkees)*, who kissed Marcia!

## What to See in Los Angeles, California

▶ **The private house at 11222 Dilling Street** (off Tujunga). This is a nice house on a quiet street near Universal City. The house appeared as the Bradys' on the show, but interestingly it is architecturally incorrect, considering the interior set of the show. The real split level house is raised on the left, not on the right, as on the set.

## FUN FACTS

- Christopher Knight, who played Peter, also played Peter's look-a-like friend, Arthur Owens.
- Carol's nephew Oliver (Robbie Rist) came for an extended visit and appeared on several episodes.
- A stage play, *The Real Live Brady Bunch* re-created the nine characters of the show using twenty-six actual scripts from *The Brady Bunch.*
- In the 1991 documentary about Kent State, *Letter to the Next Generation*, every student interviewed knew the words to "The Brady Bunch" song.
- The show's theme was sung by Peppermint Trolley Company the first season, then by *The Brady Bunch* kid actors.
- A musical comedy TV show, *The Brady Bunch Hour*, featuring the Brady characters, was on the air briefly in 1977. Guest stars on the show included Tina Turner, who sang "Rubberband Man."
- There were plans in 1971 for Carol to give birth to twins. Audience tests said a few more Bradys would be accepted, but the idea was canned.
- Actor Barry Williams became a teen idol. And due to his popularity, he was chosen to narrate *X Company*, a sea cadet filmstrip for the U.S. Navy Sea Cadet Corps. The film was aimed at boys fourteen to seventeen.
- Live-in housekeeper Alice (Ann B. Davis) earlier worked for Mike and the boys, before the families were merged.
- Robert Reed earlier appeared on *The Defenders*. Florence Henderson was a singer, who had grown up in a family of ten in Indiana. In real life she raised four kids.
- A cartoon version of *The Brady Bunch* was aired from 1972 to 1974.
- In the 1990 reunion show, *The Bradys*, Marcia has developed a bit of a drinking problem! The part of Marcia was not played by the original Marcia (Maureen McCormick—the heartthrob of many a teenage boy), but by Leah Ayres.
- *The Brady Bunch* movie, rated PG-13, debuted in 1995.

## CHARLIE'S ANGELS
**(September 1976–August 1981, ABC)**

Before *Beverly Hills 90210* and *Melrose Place*, producer Aaron Spelling found that sexy good looks and California sunshine sell, with this female detective drama. Critics called the show "jiggle TV." And there did seem to be an awful lot of bikini scenes. But that didn't stop the three angels from solving some pretty tough crimes.

The runaway star of the hit show the first year was Farrah Fawcett-Majors as Jill, she of big white teeth and blond layered hair, whose mug (and sometimes body too) graced many a magazine cover in 1976, and whose pinup poster sold more than 6 million copies.

But Fawcett-Majors left *Charlie's Angels* after a year to pursue a film career (returning for a few episodes as settlement of a contract dispute). Later, the actress criticized Spelling for not allowing the character of Jill to have a broader emotional range.

The emphasis on the good looks and good locks of the stars was no coincidence. In an *Entertainment Weekly* article in 1991, Spelling said of the show, "People would visit Universal and ask for the [show's] set—and guards would say, 'Just follow the hairbrushes.'"

On the show, the angels worked for Charlie Townsend (the voice of John Forsythe, of *Dynasty*), who was never actually seen. Charlie talked to his employees only by telephone or passed along messages through his cherubic assistant, John Bosley (David Doyle).

The original angels trained at the real Los Angeles Police Academy—at least that's the building seen on the show. *Charlie's Angels* also did location work around L.A., in addition to interior studio shots.

### WHAT TO SEE IN LOS ANGELES, CALIFORNIA

▶ **Los Angeles Police Academy,** 1880 N. Academy Drive (near Dodger Stadium). The show's original angels were all graduates of the Los Angeles Police Academy. And in the show's intro, the angels walked through the academy's real gate posts, as John Forsythe narrated "Once upon a time there were three little

The Los Angeles Police Academy. *(Photo by Fran Golden.)*

girls who went to the Police Academy . . ." In real life, there are about 360 police officers in training at the academy at any given time. The facilities are open to the public. (See also *Dragnet.*)

## Fun Facts

• The "Angels in Chains" episode has become a cult classic because more of Jill's (Farrah Fawcett-Majors) breast than usual was exposed in one scene. In the episode, the women detectives were undercover at a prison farm.

• In addition to pinup posters, Fawcett-Majors's angel popularity spurred the creation of a variety of products bearing her name, including Farrah dolls and Fawcett faucets (gold-plated and selling for $100). Her hair style started a national trend.

• Farrah Fawcett-Majors, who at the time of the show was married to Lee Majors of *The Six Million Dollar Man* (they divorced in 1982), had earlier appeared in commercials for Ultra Brite toothpaste, Noxzema skin cream, and Wella Balsam shampoo.

• After leaving the show, Fawcett gained critical acclaim as an actress off Broadway in *Extremities*, and for her TV movie role in *The Burning Bed*. On the personal front, Fawcett entered a long-term relationship with actor Ryan O'Neal, and the two had a son in 1985.

• Actress Cheryl Ladd, who replaced Fawcett as the third angel in 1977 and played Kris, was born Cheryl Stoppelmoor. She was married briefly to David Ladd, son of actor Alan Ladd. More than 3 million copies were sold of a poster of Cheryl Ladd in her angel pose.

• Jaclyn Smith, who played Kelly, earlier did commercials for Breck shampoo, Listerine mouthwash, Wella Balsam, and Max Factor. Smith later starred in several TV miniseries and lent her name to a Kmart fashion line, appearing in commercials for the discount chain. She also continued to appear on lists of the world's most beautiful women.

• Kate Jackson, who played Sabrina, later costarred in *Scarecrow and Mrs. King*. She was the angel with the most acting experience, having earlier appeared on *The Rookies* and *Dark Shadows*. She also commanded the highest angel salary.

• Shelley Hack, who played Tiffany, and who replaced Jackson as an angel in 1979, had earlier been the Charlie perfume model.

• The last of the angels, Tanya Roberts, as Julie, later appeared in the movie *Sheena*, based on the comic strip *Sheena, Queen of the Jungle*.

## CHiPs
**(September 1977–July 1983, NBC)**

This motorcycle cop show was based on real incidents involving the California Highway Patrol. The CHP cooperated in the production, and location footage for the show was shot in and around L.A., including portions of California freeways and the real CHP office in Los Angeles.

Kent Milton, a retired CHP public affairs officer, said the show always had a technical adviser from the CHP (paid for by the show). And it was the adviser's job to "make sure that uniforms, equipment, and so forth were absolutely faithful to reality." The *CHiPs* writers also had access to CHP files where they could search for ideas.

The show made a sex symbol of actor Erik Estrada, who played Francis "Ponch" Poncherello. In 1979, Estrada was seriously injured in a motorcycle accident during filming. His injuries were incorporated into the show, but contrary to popular belief, the real accident was not shown, nor was Estrada's recuperation period at the UCLA Medical Center. Rather, the hospital room was re-created on the set, after Estrada had been released from the hospital.

Estrada appeared in every episode during his recovery period via the use of old footage. He missed episodes later, however, when he temporarily left the show in a salary dispute.

## What to See in Los Angeles, California

▶ **California Highway Patrol Central Los Angeles Office,** 777 W. Washington Boulevard (under the Santa Monica Freeway). This is the main L.A. office of the California Highway Patrol, and is seen as such on the show. In addition to the exterior, *CHiPs* also filmed inside, including in the briefing room. A model of the room was later re-created for the show on a set. Visitors are welcome to tour the building.

The California Highway Patrol office in Los Angeles. *(Photo by Fran Golden.)*

## Fun Facts

• Larry Wilcox, who played Jon Baker, left the show after five seasons because he and costar Erik Estrada reportedly weren't getting along. Years later Wilcox and Estrada (their differences apparently patched up) appeared together briefly in *National Lampoon's Loaded Weapon 1.*

• Erik Estrada, who played Ponch, was born in Spanish Harlem, and earned a bit of a bad-boy reputation in Hollywood.

• Bruce Penhall, who played Bruce Nelson, was in real life a former two-time world motorcycle speedway champion.

• Tina Gayle, who played Kathy Linahan, was once a Dallas Cowboys cheerleader.

• Jon (Wilcox) and Ponch (Estrada) rode Kawasaki 1000 bikes. The real CHP has used a variety of motorcycles, with contracts usually awarded to the lowest bidder, including Kawasakis and Harley-Davidsons.

• Bruce Jenner, the former Olympic champion, appeared briefly on the show as Steve McLeish.

• In October 1981, the cast and crew of *CHiPs* donated blood during breaks in filming in honor of two real fallen CHP officers, one killed, the other injured, by gunfire.

# DRAGNET

**(December 1951–September 1959 and January 1967–September 1970, NBC)**

Starring Jack Webb as no-nonsense and highly moral cop Joe Friday, *Dragnet* is probably the most often parodied show ever on television. Mimics can't seem to resist the show's trademarks—from Friday's laconic and oft-repeated "Just the facts, ma'am," to the mindless head-nodding the cops seem to do whenever they figure out a crime.

The show was ground-breaking in its first run for its efforts at documentary-style realism and its lack of slapstick comedy, or any humor, for that matter (although some would argue Webb's exaggerated portrayal of a tough plainclothes cop was pretty funny).

*Dragnet* was so popular it ranked as the number two show in 1953, after *I Love Lucy,* and was the third highest rated show in 1954, after *I Love Lucy* and *The Jackie Gleason Show.*

The show was shot in Hollywood, and featured some real locations including the real headquarters of the Los Angeles Police Department and the real Los Angeles City Hall building.

Friday went through several partners, but Officer Frank Smith (Ben Alexander) was his partner for much of the 1950s.

Probably the best recognized of Friday's partners was Harry Morgan *(M*A*S*H),* who played Officer Bill Gannon. It was Gannon who helped Friday combat mostly hippie-types and drugs in the 1960s.

## What to See in Los Angeles, California

▶ **Parker Center,** 150 N. Los Angeles Street (near City Hall). The real headquarters of the Los Angeles Police Department appeared as itself on the show. Only the reception desk is open to visitors, but if you hang around outside long enough you're bound to spot some plainclothes cops.

The Parker Center, headquarters of the L.A.P.D. *(Photo by Fran Golden.)*

▶ **Los Angeles City Hall,** 200 N. Spring Street. The real city hall appeared as itself on *Dragnet.*

▶ **Los Angeles Police Academy,** 1880 N. Academy Drive (near Dodger Stadium). The badge used by Joe Friday (Jack Webb), number 714, is on display in the academy's cafe. And actor Webb's personal guns are in a case in the Police Training and Recreation Center building. *Charlie's Angels* showed the academy's main gate in its intro. The academy facilities, built in the 1930s, are open weekdays to the public (weddings are even held in the rock garden area). The cafe is open Monday to Friday, 6 A.M. to 2 P.M.

▶ **The Los Angeles Police Revolver and Athletic Club Store,** at the Los Angeles Police Academy. The store offers L.A.P.D. T-shirts and souvenirs including photos of the cast of the 1987 *Dragnet* film starring Dan Aykroyd and Tom Hanks. Open Monday to Thursday, 7:30 A.M. to 4 P.M., and Friday until 5 P.M. Call 213-221-3101.

## Fun Facts

• "It's my job, ma'am," character Friday would stiffly say. Jack Webb's job on the show included acting, directing, and producing.

• Jack Webb also produced other reality-based shows including *Adam 12* and *Emergency!*

• While Joe Friday was a bachelor, actor Webb had four wives, including one who was a former Miss U.S.A.

• The first notes of the *Dragnet* theme song, "dummm-dee dum-dum," became among the most recognizable four notes ever. The song, written by Walter Schumann, was a hit in the 1950s.

• Jack Webb's earlier roles included a small part in the campy movie classic *Sunset Boulevard.*

• Webb's efforts at realism were such that he once reportedly gathered cigarette butts from L.A.P.D. headquarters to use in the ashtrays on the *Dragnet* set.

• George Fenneman was one of the announcers on the show. He was also Groucho Marx's sidekick on *You Bet Your Life.*

• When Jack Webb died of a heart attack in 1982, flags at the L.A.P.D. headquarters were flown at half-staff.

• Reruns of the show ran under the title *Badge 714.*

## L.A. LAW
**(October 1986–May 1994, NBC)**

Sorry, but Stuart Markowitz's (Michael Tucker) special sex ritual, the Venus Butterfly, is fictional.

But as farfetched as they sometimes seemed, many of the cases and situations on this much-talked-about show were inspired by real newspaper headlines. The show introduced many TV viewers to real Tourette's syndrome, for instance. In a memorable case, a businessman client, suffering from the ailment, loses his job because he is unable to stop blurting out obscenities.

The show had such an impact on American jurisprudence that its well-written cases even found their way into real courtroom debates.

But not all cases were serious and not all the lawyers won. And there were soap opera elements to the show too, including office politics and love affairs.

The Jaguar XJ6 seen in the opening credits, with the "LA LAW" license plate, set the tone for the *L.A. Law* lawyers, whose private, office, and courtroom lives were detailed on the show. The *L.A. Law* lawyers are a successful bunch, earning top dollar—Arnie Becker's (Corbin Bernsen) salary is mentioned in a 1994 episode as being in the high six figures—and they have very comfortable, and sometimes glitzy, life-styles that, in several cases, include lots of sex (Stuart with Ann, Grace with Michael, Grace with Victor, Arnie with anyone).

Some real lawyers complained that the *L.A. Law* crew was more eccentric and corrupt than in real life. But hey, this is L.A.

And the sometimes zaniness of the show did not stop potential law students from swarming law schools with applications, especially as the show increased in popularity.

*L.A. Law* was created by Steven Bochco, who had earlier gained fame and critical applause for *Hill Street Blues*, and Terry Louise Fisher, a former deputy DA, who had been a producer for *Cagney & Lacey*.

The show created backstage legal headlines itself when Fisher and Bochco ended up breaking off their business relationship in 1987, with Fisher getting a settlement in a resulting $50 million lawsuit the following year.

## What to See in Los Angeles, California

- **The Office Building at 444 S. Flower Street** (next to the L.A. Public Library). This downtown high rise was the home of McKenzie, Brackman et al. on the show and is in real life home to law offices and other businesses. In the city, it is known as the *L.A. Law* building.

- **Tripp's Restaurant,** 10131 Constellation Boulevard (in Century City). This two-level restaurant (one level is a balcony) was frequently seen on *L.A. Law.* The crime show, *Columbo,* also used the location. The cuisine is continental, with an emphasis on steaks. The restaurant is open for lunch, Monday through Friday, and for dinner, Thursday through Saturday. Call 310-553-6000.

The *L.A. Law* building is the one on the right. *(Photo by Fran Golden.)*

## Fun Facts

• While his character, Michael Kuzak, liked to date lawyers, hunky Harry Hamlin liked actresses. He married and divorced Laura Johnson of *Falcon Crest* and Nicollette Sheridan of *Knot's Landing*, and later dated Lisa Rinna of *Days of Our Lives*.

• McKenzie, Brackman partner Chaney, one of the founders of the firm, died at his desk at the beginning of the series's pilot. In response to his death, a compassionate Arnold Becker (Corbin Bernsen) said, "I've got dibs on his office."

• *L.A. Law* was somewhat of a family affair. Alan Rachins, who played Douglas Brackman, Jr., is married in real life to actress Joanna Frank, who also played his on-screen wife. She is series cocreator Steven Bochco's sister.

• Also married in real life are Michael Tucker, who played Stuart Markowitz, and Jill Eikenberry, who played Ann Kelsey.

• Actor Corbin Bernsen's real-life mom, Jeanne Cooper, of *The Young and the Restless*, played character Arnie's mom on the show.

• A poll of L.A. attorneys in 1991 by *Crime Beat Magazine* showed Victor Sifuentes (Jimmy Smits) as the character the lawyers would choose to represent them if they needed a lawyer. But the lawyers said the most realistic character was Douglas Brackman (Alan Rachins).

• Jimmy Smits later played a cop on *NYPD Blue*.

• Michele Greene, who played Abby Perkins, said repeatedly in interviews that she did not leave the series because of the lesbian storyline involving her character and C. J. Lamb (Amanda Donohoe). Greene said she left because she wanted to pursue other interests including her singing career.

• Diana Muldaur, who played Rosalind Shays, used a double for the empty elevator shaft plunge scene. The character Shays died in the fall.

• In a 1992 article in *The New York Times*, Stephen Gillers, a professor of legal ethics at New York University School of Law, said *L.A. Law* invokes the most exciting legal issue discussions among students at the school.

• Susan Dey, who played Grace Van Owen, was most famous earlier for her role as Laurie on *The Partridge Family*. She was the only member of the original *L.A. Law* cast not to appear in the pilot.

• In an interview with the Associated Press during filming of the show's final episode, Corbin Bernsen, a married family man with three kids, waxed nostalgic, saying the sofa in character Arnie's office makes him think, "That girl, this girl, that girl, this girl."

• Steven Bochco's protégé, David E. Kelley, a Boston University Law School grad who had submitted a script to Bochco and was promptly hired as one of the show's main writers, took over as executive producer in 1989. Kelley himself left two years later to develop *Picket Fences.*

• In an unusual move, *L.A. Law* was pulled off the air for a month in 1993 for retooling, after its ratings declined. Viewers were advised in an on-air announcement that the *L.A. Law* they remembered would be back. It came back, but only for a year.

• *The New York Times* reported in May 1990 that a Miami medical malpractice case was so similar to an *L.A. Law* script that a lawyer asked a judge to question jurors as to whether they had seen the show. The motion was denied.

## LOU GRANT
**(September 1977–September 1982, CBS)**

When gruff but warm-hearted Lou Grant (Ed Asner) got fired from his TV job at WJM in Minneapolis, on *The Mary Tyler Moore Show,* he looked up his old friend Charlie Hume (Mason Adams), managing editor of the *Los Angeles Tribune.* And Charlie offered Lou a job as city editor. The only catch was that Lou had to learn to get along with Mrs. Margaret Pynchon (Nancy Marchand), the owner/publisher of the *Tribune,* who could be just as tough (but also just as compassionate) as Lou.

The producers sought to show the quality drama's fictional newsroom in a realistic light, with the characters debating real ethical issues, racing to meet deadlines, jockeying for front-page bylines, and even striking.

The show created headlines of its own when it was canceled in 1982, with charges that the reason was not just low ratings but the liberal politics of its star. Ed Asner was at the time involved with efforts to get medical relief to victims of the war in El Salvador, and was the outspoken president of the Screen Actors Guild.

## WHAT TO SEE IN LOS ANGELES, CALIFORNIA

- **The Title Guarantee and Trust Building,** 411 W. 5th Street (downtown). This art deco-style building really houses a newspaper among its tenants. But it is the Spanish-language *La Opinion,* and not the fictional *Los Angeles Tribune.*

The Title Guarantee and Trust Building doubled for the headquarters of the *Los Angeles Tribune* on *Lou Grant.* *(Photo by Fran Golden.)*

## Fun Facts

- Reverend Jesse Jackson appeared on *Lou Grant* in 1978, as himself. He spoke at a ghetto high school, telling kids about the importance of a good education.
- Real L.A. news photographers complained that the show's hippy photographer character, Animal (Daryl Anderson), was not a true representation of their profession.
- Robert Walden, who played ambitious journalist Joe Rossi on the show, was in the movie *All the President's Men* playing one of the bad guys, not a journalist.
- Ed Asner won three Emmy awards for his role as Lou Grant on *The Mary Tyler Moore Show,* and two for his role as the same newsman on *Lou Grant.* The actor also won Emmy awards for his roles in the miniseries *Rich Man, Poor Man* and *Roots.*

## THE LOVE BOAT
**(September 1977–September 1986, ABC)**

This hit comedy show featured multiple vignettes each week about love and romance. The show literally became a prime-time commercial for the cruise industry. Millions of people, in ninety-three countries, tuned in to see Captain Merrill Stubing (Gavin MacLeod) and his crew, as well as a new cast of guest stars each week, aboard the real luxury cruise ship the *Pacific Princess* (or its twin, the *Island Princess*), as it cruised to Mexico, the Caribbean, or Alaska. Los Angeles-based Princess Cruises, which owned the ships, hired actor MacLeod as its commercial pitchman, further assuring a tie-in in the minds of viewers.

Real passengers on the ships during the show's run were cast as extras. And voyages on the *Pacific Princess* and *Island Princess* on which filming for *The Love Boat* took place were often sold out well in advance.

*The Love Boat,* with its endless array of happy endings, of course added a Hollywood slant to the cruise experience (the crew and guests don't often interact that much in real life), but viewers saw enough to know that cruising can be fun, and not just for white-gloved old ladies.

And some things depicted on the show really do happen. Captain Stubing met and fell in love on the ship with Emily (Marion Ross), whom he married. And in real life Captain John Crichton, of the real *Pacific Princess*, also met his wife at sea, *Travel Weekly*, the travel industry newspaper, reported.

Some of the show's interiors were shot on a Hollywood set, including the cabin scenes (the cabins shown on the show were larger than on the real ships).

*The Love Boat* also filmed aboard other cruise ships including Sun Line's *Stella Solaris*, which was seen on a series of specials featuring a cruise in the Mediterranean, and Pearl Cruises's *Pearl of Scandinavia*, which was seen on a special featuring a cruise with stops in China, Hong Kong, and Japan. A reunion show in 1989 was shot in the Caribbean aboard the *Sky Princess*, another Princess Cruises vessel.

The pilot for *The Love Boat* was shot aboard a ship called the *Sun Princess*, which has since retired (a new ship will bear that

Princess Cruises' *Pacific Princess* was made famous as *The Love Boat*. It's shown here in the Caribbean. *(Photo courtesy of Princess Cruises.)*

name starting in 1996). But both the *Pacific Princess* and *Island Princess* continue in service and were recently renovated (at a cost of $40 million), to offer passengers an even more glamorous cruise setting.

## What to See in Los Angeles, California

▶ **The Pacific Princess and the Island Princess.** Part of L.A.-based Princess Cruises's fleet of nine ships, the two five-star-rated ships that appeared on the show are both 20,000-ton vessels that carry 640 passengers. Each has a two-tiered lobby with a spiral staircase, lounges with floor-to-ceiling windows, cozy bars, and a casino. The ships each have a fully-equipped gym, two swimming pools, a jogging track, and sunbathing areas. For those who can't do without, each stateroom is equipped with its own color TV. The ships are mostly used these days for Asia, Europe, and other exotic routes, but also once a year or so visit L.A. For reservations call 800-LOVE-BOAT.

## What to See in San Pedro, California

▶ **The Port of Los Angeles,** Los Angeles Harbor. Princess Cruises ships can be viewed here from October through early May. Most often docked at the port are the *Fair Princess* and the *Golden Princess*.

## Fun Facts

- Gavin MacLeod was earlier Murray on *The Mary Tyler Moore Show,* and also appeared on *McHale's Navy.*
- The theme song for the show was written by Paul Williams and Charles Fox.
- Fred Grandy, who played assistant purser Gopher, became a congressman from Iowa.
- Charo appeared on *The Love Boat* as entertainer April Lopez.
- The show was based on *The Love Boat,* a book by Jeraldine Saunders, which described her life as a cruise director.
- The series was translated into twenty-nine languages for foreign viewing.

• Famous movie, stage, and TV stars appeared as guests on *The Love Boat* and got a cruise out of the deal. They included Helen Hayes and Raymond Burr *(Perry Mason, Ironside).*

• If the show looked like an expanded, floating version of *Love, American Style,* it's because *The Love Boat* was created by the same man, producer Douglas Cramer.

## MELROSE PLACE
**(July 1992– , Fox)**

This campy show started off as a serious drama; a *thirtysomething* for twentysomethings. But the format wasn't working. The early episodes gained some attention on college campuses, where students would gather together to guffaw as former soccer pro Andrew Shue, who plays Billy Campbell, tried to act (often by rolling his eyes). But *Melrose Place* did not have broader appeal.

So producer Darren Star, a thirtysomething himself, decided to convert the show into a soap opera, with lots of sex. And he brought in heavy-hitter Heather Locklear to lead the charge. Locklear, who had played Sammy Jo, Alexis's protégée on *Dynasty* (like *Melrose Place,* an Aaron Spelling production), brings a delightful bitchiness to her role as Amanda, the up-and-coming advertising executive, apartment building owner, and seductress (supposedly, Locklear is in real life a really nice person). The actress is credited with saving the show from oblivion. And her cat fights with the good girl of the show, Allison, provide plenty of coffee-hour fodder.

Other topics for discussion among fans are story lines involving a Hollywood call girl ring, a DUI accident, blackmail, Russian émigrés, a lonely computer millionaire, and even a return from the dead.

The show, and its gorgeous stars (both male and female) gets lots of ink in newspapers and magazines around the country, and supermodel Cindy Crawford even did a special visit to the set of *Melrose Place* on her MTV fashion show (how much more trendy can you get?). Off-screen romances among the cast also have provided tidbits for the tabloids and fan magazines.

In real life there is a Melrose Avenue in L.A., and it is the city's trendiest shopping and cafe area, attracting plenty of members of the Generation X crowd. Shots of the area are seen in the show's intro. And there is also a real street called Melrose Place, with street signs under which fans can pose. But, alas, there is no real stucco apartment building with a swimming pool on the street.

The Melrose Place of the show exists only on a set in Santa Clarita, about thirty-five miles north of Hollywood, that is closed to the public. Shooters, the bar where the characters sometimes hang out, also exists only on the set, a spokeswoman for the show's production company said.

## What to See in Los Angeles, California

▶ **Melrose Place** (off La Cienaga Boulevard and Melrose Avenue). The real street is inhabited by upscale antique shops that twentysomethings probably can't afford. There is no 4616 Melrose Place, the address of the apartment complex on the show. But there are street signs that say Melrose Place.

▶ **Melrose Avenue** (in particular, between La Brea and Fairfax). This trendy area offers plenty of shopping and entertainment

The real Melrose Place in L.A. *(Photo by Fran Golden.)*

options for twentysomethings. The Burger That Ate L.A., the restaurant that appeared in the show's intro, was located at 7624 Melrose. But it was later converted to a Mexican chicken restaurant.

## Fun Facts

• *Melrose Place* is a spin-off of *Beverly Hills 90210*.

• Heather Locklear was once married to Tommy Lee, the drummer of the heavy metal rock band Mötley Crue. Locklear was originally signed to appear in only four episodes of *Melrose Place*. In 1994 she married Richie Sambora, lead guitarist for Bon Jovi.

• Courtney Thorne-Smith, who plays Allison, described herself to *TV Guide* as being "funny, warm, and geeky." She earlier appeared as a Lakers cheerleader on *L.A. Law*.

• The apartments at the show's fictional Melrose Place complex are one or two bedrooms, and rent for about $800 a month.

• Grant Show, who plays hunky Jake Hanson, was earlier on *Ryan's Hope*. Jake was first introduced on *Beverly Hills 90210* as the temporary love interest of Kelly (Jennie Garth).

• In real life, Grant Show dates red-haired Laura Leighton, who plays psychotic Sydney. And Andrew Shue, who plays Billy, dated Courtney Thorne-Smith, who plays his on-again, off-again on-screen lover, Allison. (Billy marries conniving Brooke, played by Kristin Davis, in a May 1995 episode.)

• Doug Savant, who plays the gay character of Matt Fielding, is in real life a married father of two.

• Josie Bissett, who plays Jane, is married in real life to Rob Estes, star of USA Channel's *Silk Stalkings*.

• Andrew Shue, a graduate of Dartmouth College, is the brother of actress Elizabeth Shue. He played professional soccer in Zimbabwe and used his fame on *Melrose Place* to create Do Something, a nonprofit organization offering grants to those with ideas to help solve social problems.

• Heather Locklear joked with Cindy Crawford on MTV that since character Amanda has had most of the men at Melrose Place (she has slept with Billy and Jake), she should take a female lover.

## REMINGTON STEELE
**(October 1982–August 1986, NBC)**

Stephanie Zimbalist starred in this detective show as Laura Holt, a brassy female private detective who set up her own agency, Laura Holt Investigations, but found her clients really wanted to deal with a man. Holt invented Remington Steele, as her handsome and sophisticated fictional boss, and changed the name of the agency to Remington Steele Investigations.

Business was going well when a dashing Irish ex-jewelry-thief entered the picture and, looking for a cover, offered to become the real Remington Steele (Pierce Brosnan). Laura gained a partner, and a love interest, too.

The show was designed to play like a 1940s movie, with romantic tension and even hand-kissing. Remington Steele (Brosnan) was an old movie fan who relied on the plots of such classics as *The Thin Man* to solve cases.

The show was filmed at Studio City, with the building shown as the home of Remington Steele Investigations actually the high-rise twin Century Plaza Towers in L.A. *Remington Steele* also went on location to Ireland, Manzanillo, Mexico, Malta, and Cannes, France.

### What to See in Los Angeles, California

▶ **Century Plaza Towers,** 2029 and 2049 Century Park East (off Santa Monica Boulevard). The tall office towers, which were the location of Remington Steele Investigations on the show, have also appeared on *Melrose Place.*

### Fun Facts

• Stephanie Zimbalist is the daughter of Efrem Zimbalist, Jr. *(The F.B.I.).* He guest-starred on the show as a con artist who taught Remington Steele (Pierce Brosnan) the fine art of thievery.

• Laura Holt (Zimbalist) and Remington Steele (Brosnan) are married on the show in 1986, but only so he won't be deported to Ireland. The two do finally consummate their relationship in a 1987 TV movie, however.

The Century Plaza Towers.
*(Photo provided by The Century Plaza Towers.)*

- Henry Mancini wrote the show's theme song, "Remington Steele."
- Actor Brosnan, who is from County Meath, Ireland, was talked about to replace Roger Moore as the next James Bond in the mid-1980s, but he was tied up with the show, and Timothy Dalton got the movie part. Brosnan did eventually get to play agent 007, however, in the 1995 James Bond movie *Goldeneye.*

## FYI: ALSO IN LOS ANGELES

The home of the Nelson family on the long-running family sitcom *The Adventures of Ozzie and Harriet* (October 1952 to

September 1966, ABC) was modeled after the real-life home of the show's stars, Ozzie and Harriet Nelson, who lived in Hollywood.

### What to See in Los Angeles, California

▶ **The former home of Ozzie and Harriet Nelson,** 1822 Camino Palmero Drive. This is where the Nelson family, Ozzie, Harriet, David, and Ricky lived in real life.

On *The Beverly Hillbillies* (September 1962 to September 1971, CBS), Beverly Hills banker Milburn Drysdale (Raymond Bailey) worked out of the Commerce Bank of Beverly Hills.

A real building was seen on the popular sitcom as the Commerce Bank, but it's not in Beverly Hills, and not a bank at all, but rather an L.A. office building.

### What to See in Los Angeles

▶ **The Carnation Milk Building,** 5055 Wilshire Boulevard (near Highland Avenue). Seen as the Commerce Bank on the show, this building, which once housed the administrative offices for Carnation Milk, today has among its tenants the *Hollywood Reporter.*

(See also Los Angeles County and Environs)

When the kids from West Beverly High graduate on *Beverly Hills 90210* (October 1990–, Fox) they go off to college close to home, at California University. There is no real California University, but there is a real college campus that is seen on the show. It's Occidental College, a four-year liberal arts school with a student population of 1,650, located in L.A.

### What to See in Los Angeles

▶ **Occidental College,** 1600 Campus Road (in the northeast portion of L.A., near Pasadena and Glendale). This real-life college serves as the location for the fictional college on the show. Exterior shots shown frequently include the Main Quad area in front of the Freeman Student Union. The Thorne Hall Theatre is seen in interior and exterior shots as the place where Brenda performs in a play. The Faculty Club, on the north end of the campus, is seen as Steve's Fraternity, the KEG House (Kappa Epsilon Gama) on the show. The dormitory where Andrea lives is really Occidental's Haines Hall, although interior shots were done in a re-creation of a dorm room on the show's set. The

The Thorne Hall Theatre at Occidental College. *(Photo courtesy of Occidental College, Los Angeles.)*

The Freeman Student Union at Occidental College. *(Photo courtesy of Occidental College, Los Angeles.)*

show also filmed in the college's library, and the real-life Fowler Science Hall was the setting on the show for a sixties-style sit-in.
(See also Los Angeles County and Environs)

The setting for *Happy Days* (January 1974 to July 1984, ABC) was Milwaukee, Wisconsin, but a real private house was shown as the Cunningham family's, and it is in Hollywood, not far from the studio where interiors for the show were filmed.

### What to See in Los Angeles

▶ **The private house at 565 N. Cahuenga Boulevard** (near Paramount Studios, in Hollywood). This private house on a quiet side street doubled for the Cunningham's Milwaukee home on the show. The midwestern-style architecture stands out in this southern California neighborhood.
(See also The Midwest)

The sitcom *It's a Living/Making a Living* (October 1980 to September 1982, ABC, and 1985 to 1989, syndicated) took place in the posh Above the Top restaurant on the thirtieth floor of an L.A. high rise. In real life, the building shown is the Westin

The *Happy Days* house in Hollywood. *(Photo by Fran Golden.)*

The Westin Bonaventure Hotel. *(Photo courtesy of the Westin Bonaventure Hotel, Los Angeles.)*

Bonaventure Hotel, a massive futuristic-looking downtown hotel with glass elevators that was once named one of the ten most photographed buildings in the world by *Fortune Magazine*.

## What to See in Los Angeles

▶ **Westin Bonaventure Hotel, Top of Five Restaurant,** 404 S. Figueroa Street (downtown). The hotel's thirty-fifth-floor fine-dining restaurant was the inspiration for the show's Above the Top. The restaurant offers celebrated views of L.A., Santa Monica, and the Pacific Ocean, and features fresh American cuisine. The restaurant is open for dinner, Sunday through Thursday, 5:30 P.M. to 10 P.M., and Friday and Saturday, 5:30 P.M. to 11 P.M. Lunch is served on weekdays, 11:30 A.M. to 2:30 P.M. Built in 1976, the hotel itself has 1,368 guest rooms, and has also appeared on *L.A. Law* and *Moonlighting*, and in numerous movies. Reservations are suggested. Call 213-624-1000.

# LOS ANGELES COUNTY AND ENVIRONS

When they head outside the city for location shoots, Hollywood production crews don't always go very far. For Philadelphia-based *thirtysomething,* they headed to South Pasadena for a shot of a funky house to call Hope and Michael's. And the exotic setting for *Fantasy Island* was really in Arcadia. Many of the legendary southern California movie ranches have, unfortunately, given way to suburban expansion, including the old Iverson's Ranch in the San Fernando Valley, which was used as a location for *Bonanza, Gunsmoke,* and *The Big Valley.*

But a few are now open to the public.

The Twentieth Century-Fox ranch, used for scenes in *M*A*S*H,* for instance, is operated as a state park. And visitors can tour the *M*A*S*H* location site (where the choppers landed) and see props left behind by the show. Visitors can also visit Paramount Ranch, now operated by the National Park Service, and used by *Dr. Quinn, Medicine Woman* as a setting.

Southern California's gorgeous natural beach scenery has been featured in many TV shows, most prominently in *Baywatch,* which literally shows locations up and down the coast.

On the Santa Monica Beach, the historic Santa Monica Pier, with its antique carousel, appeared in at least one episode of *The Mod Squad,* and more recently was seen in *Seinfeld* (where it was supposed to be in New York), *Beverly Hills 90210, Baywatch,* and *Melrose Place.*

*Beverly Hills 90210* also shot scenes at a former beach club on Santa Monica Beach (Palisades Beach Road), but the location was closed after the 1994 L.A. earthquake. *Beverly Hills 90210* actually did film a few scenes in Beverly Hills, but the house seen as the Walsh family's is in Altadena, California.

The sitcom *Three's Company* in its intro showed Jack, Janet, and Chrissy riding bikes on Ocean Front Walk in Venice, California. And on *The Rockford Files*, Jim Rockford's trailer really was parked at the beach in Malibu.

Real suburban high schools were seen on *Beverly Hills 90210*, although none is actually in Beverly Hills. Suburban high schools were also seen on other teen shows including *The Wonder Years* and *Life Goes On*, often with real students and teachers in the background.

Also using Beverly Hills as a location site is *NYPD Blue*, which featured the presidential suite at the very posh Regent Beverly Wilshire in one episode (the setting was supposed to be Manhattan). A Beverly Hills Greek restaurant, The Greek Connection, also appeared on the New York cop show.

*The Beverly Hillbillies* showed a touch of the real Beverly Hills. And so did *I Love Lucy*, with Lucille Ball's real home appearing on the show. *Here's Lucy* (Ball's later show) filmed some scenes on location, including a hilarious episode that featured the real Los Angeles International Airport. On the show, Lucille Carter took what one writer called "a wild, wacky ride on a luggage belt."

*The Partridge Family* did some location work too, visiting amusement parks and the like for performances (including one in Ohio), but there is no real San Pueblo, California, the hometown of the musical family on the show. *The Partridge Family* was shot mostly on a Hollywood set.

## Baywatch

**(September 1989–August 1990, NBC, then syndicated, September 1991– )**

Buff bodies and bikini-clad babes are featured in this California beach lifeguard show, which also shows off the natural scenery of the southern California coastline. A much bigger hit in syndication than in its original run, the show has earned the nickname *Babewatch* from guys, for its frequent bikini shots, and *Bodywatch* from gals, for its frequent muscle shots.

But the fictional L.A. County lifeguards on the show's Malibu Beach do more than sit around showing off their tans and looking at the Pacific Ocean. They have daring rescues, robberies, rapes, custody battles, love affairs, and other assorted dramatic crises to deal with.

With all its outdoor footage and location shoots on real California beaches, *Baywatch* was originally very expensive to produce, costing about $1.2 million an episode, and NBC canceled it after a year. But series star David Hasselhoff *(Knight Rider)* took over as co–executive producer and got new investors interested, cutting costs a bit by filming more of the action in a sound studio in Culver City. *Baywatch* is a top show in its syndicated form, particularly popular in the United Kingdom, Australia, Holland, and Germany.

## WHAT TO SEE IN SANTA MONICA, CALIFORNIA

- **The lifeguard station at Will Rogers State Beach,** on the Pacific Coast Highway at Temescal Canyon. This is the real lifeguard station seen on the show, and beach scenes are filmed here as well. But other beaches, from Orange County to Ventura County, including nearby Santa Monica State Beach, are also featured, according to a crew member of the show.

## FUN FACTS

- Pamela Denise Anderson (C. J. Parker) was earlier Lisa, the "Tool Time girl" on *Home Improvement.* A former Playboy Bunny, she was once engaged to actor Scott Baio *(Happy Days),* whom she met at the Playboy Mansion in 1989.
- Anderson also reportedly briefly dated hunky costar David Charvet (Matt Brody), who was once a Bugle Boy jeans model.
- In 1995, Anderson married Tommy Lee, Heather Locklear's *(Melrose Place)* ex.
- David Hasselhoff, who plays Lieutenant Mitch Bucannon, is also a singer. His albums and performances are particularly popular in Europe and Japan.

The lifeguard station at Will Rogers State Beach, as seen on *Baywatch*. *(Photo by Fran Golden.)*

• Actress Pamela Bach, who appears on *Baywatch* as reporter Kay Morgan, is the real Mrs. David Hasselhoff.

• Gregory Alan-Williams, who plays Garner Ellerbee, became a hero in real life during the 1992 riots in L.A. that followed the not-guilty verdict in the police beating of Rodney King. Alan-Williams stepped in to save the life of a Japanese-American man who was being beaten in front of two hundred onlookers.

• Before *Knight Rider*, David Hasselhoff starred on the soap *The Young and the Restless*.

## THE BEVERLY HILLBILLIES
**(September 1962–September 1971, CBS)**

Jed Clampett (Buddy Ebsen) was a poor Ozark mountaineer *"who barely kept his family fed."* And then one day he was out shooting for food and hit pay dirt instead, in the form of *"bubbly crude. Oil that is."*

Widower Jed takes the millions he makes selling his oil-rich land to the O.K. Oil Company and heads to Beverly Hills. He is joined by his spunky elderly mother-in-law, Granny (Irene Ryan), his blond unmarried daughter, Elly Mae (Donna Douglas), and his big dense nephew, Jethro Bodine (Max Baer, Jr.).

The rest is television history.

Although *The Beverly Hillbillies* was the number one show in the 1963 season, and again in 1964, and one of the top comedies of the 1960s, the critics, of course, hated the show, calling its humor lowbrow.

*The Beverly Hillbillies* was shot mostly in a studio, but exteriors of a real Bel-Air mansion doubled for Jed's place on the show. Unfortunately, the real mansion was dismantled in the 1980s, when its owner, a Hollywood agent/producer, built an even larger place on the property at 750 Bel Air Road. (Interestingly, the property is right next door to the home where Ron and Nancy Reagan retired).

The hillbillies were seen driving their old jalopy on real palm-tree lined streets in posh Beverly Hills. And the show went on location several times, including to Branson, Missouri (for Ozark scenes), Washington, D.C., and New York.

The cast and crew also went to England in 1967, filming at Penshurst, a real castle that's more than six centuries old located about forty miles from London, in Tonbridge, Kent.

The cast occasionally went "back home" to visit friends in Hooterville, the fictional setting for sister shows *Petticoat Junction* and *Green Acres*, as well.

All three rural comedy shows were developed by writer/producer Paul Henning, who is originally from Missouri.

## What to See in Beverly Hills, California

▶ **Beverly Drive** (the road cuts north to south through Beverly Hills). Palm-tree-lined parts of this roadway were seen when the Clampetts drove around Beverly Hills in their rickety car, according to series creator Paul Henning. While the car was occasionally taken out on the real road, most outdoor footage was filmed without the vehicle. The cast did their scenes in the car, in the studio, with the background shots added through a rear projection process, Henning said.

## FUN FACTS

• Buddy Ebsen was the original actor cast as the Tin Man in *The Wizard of Oz*, but he had a serious allergic reaction to the silver makeup and was replaced by Jack Haley, Sr.

• After *The Beverly Hillbillies*, Ebsen played a detective on *Barnaby Jones*. He made a cameo appearance as the detective character in the 1993 movie *The Beverly Hillbillies*.

• Actor Ebsen is also a noted yachtsman, who has won several races. He served in the Coast Guard in World War II.

• In a 1967 bullfighting scene, Jethro (Max Baer, Jr.) wore a matador suit that was originally custom made for Anthony Quinn for *The Magnificent Matador.* Actor Baer was 6 foot, 4 inches tall at the time, and weighed 215 pounds, so finding a costume to fit was a bit of a task.

• Irene Ryan said fans often called her by her character's name, Granny, but she didn't mind. "Heck, I'd be tickled if folks called me that for the rest of my life," she said in a CBS news release. They practically did. The show completed production in 1971, and Ryan passed away in 1973.

• Granny's first name is Daisy.

• The role of Commerce Bank secretary Janet Trego was played by actress Sharon Tate, who was later among the victims of mass murderer Charles Manson.

• In *The Return of the Beverly Hillbillies* special aired in 1981, Miss Hathaway (Nancy Kulp) worked for the Department of Energy and Elly Mae (Donna Douglas), who loved "critters," ran Elly's Zoo.

• "The Ballad of Jed Clampett" became a hit record in 1963. The song was sung by Jerry Scoggins, and written and played by Lester Flatt and Earl Scruggs, who also occasionally appeared on the show.

• In addition to playing Jethro, Max Baer, Jr., also played the part of Jethro's sister, Jethrene.

• Paul Henning, the man behind *The Beverly Hillbillies* and the subsequent rural comedies *Green Acres* and *Petticoat Junction* was earlier a writer on *The George Burns and Gracie Allen Show.*

• *The Beverly Hillbillies* featured comedians in guest appearances, including Soupy Sales, Phil Silvers, and Paul Lynde.

## BEVERLY HILLS 90210
**(October 1990– , Fox)**

This young people's drama has produced some of America's most popular teens (although some of the actors were really past their teens when the show began).

And so familiar have they become that viewers don't seem to mind at all that the Beverly Hills gang graduated from high school, and the 90210 zip code, and has moved on to college.

Twins Brenda (Shannen Doherty) and Brandon (Jason Priestley) Walsh were sixteen, and fresh from Minnesota, when the show began, having followed their accountant dad Jim's (James Eckhouse) career path out to posh Beverly Hills.

The Walsh twins quickly found friendships with the mostly rich kids from broken homes at West Beverly High. They adjusted to wearing great clothes and exhibiting mature fashion know-how, while mom Cindy (Carol Potter) provided comfort food from their nonobtrusive, comfortable Beverly Hills house, for the masses.

The show deals with real young adult issues such as sex, pregnancy, drugs, alcohol, eating disorders, divorce, and mental illness. And it set a fashion standard, for a time, in Brandon's (Priestley) and Dylan's (Luke Perry) sideburns. By 1993, the show was watched by an estimated 200 million people around the world.

The off-the-set life of tempestuous actress Shannen Doherty also created plenty of fodder for the tabloids. And several of the show's stars, including Priestley and Perry, have springboarded from their popular TV roles into movie parts. Jennie Garth, who plays Kelly, also has had starring roles in several TV movies.

The house seen on the show as the Walsh's is a real house, but not in Beverly Hills. And neither is the real high school that doubles for West Beverly High. The beach house initially shared by Donna (Tori Spelling), David (Brian Austin Green), and Kelly (Garth) is really in the seaside town of Hermosa Beach (about fifteen miles south of L.A.). Or at least it was until residents kicked the show out of town, saying the filming destroyed the town's tranquility.

According to Beverly Hills officials, the show has shot some footage in the city, including shots of the real Roxbury Park, a public park, the Beverly Hills Civic Center, and a nightclub called Tattoos (at 233 North Beverly Drive).

## What to See in Altadena, California

▶ **The private house at 1675 E. Altadena Drive** (off Lake Avenue). This house, which is nowhere near Beverly Hills (it's really north of Pasadena), was seen as the Walsh family's house on the show.

## What to See in Torrance, California

▶ **Torrance High School,** 2200 West Carson Street. This real high school was seen as West Beverly High School on the show. The main building of the school was built in 1917, and is a historical landmark. The campus itself is a half-mile long, is made up of several buildings, and has a student population of 1,750. There are really outside eating areas, as frequently seen on the show, although for the show the outdoor tables were moved to locations not normally used (but which made for better shots). Crews also

The house seen as the Walsh family's on *Beverly Hills 90210. (Photo by Fran Golden.)*

filmed inside the school, but interiors were also done at other area schools, as well as on a classroom set. The zip code at the school is really 90501.

## What to See in Van Nuys, California

▶ **Grant High School,** 13000 Oxnard Street. This high school was used for some interiors on *Beverly Hills 90210.* It also appeared on *The Wonder Years, Life Goes On,* and *Parker Lewis Can't Lose.*

## Fun Facts

• Jason Priestley also does commercials for Pepe jeans.

• Tori Spelling, who plays Donna, and is the daughter of the show's executive producer, Aaron Spelling, lived with her family in a 123-room mansion, but is said to be not at all pretentious in real life (although she reportedly loves to go shopping). On the love front, she hung for a while with Nicholas Savalas, son of Telly Savalas *(Kojak).*

• Shannen Doherty's good-girl character spawned an "I Hate Brenda Newsletter," but in real life, Doherty made headlines for bounced checks and barroom fights.

• Ian Ziering, Luke Perry, and Jason Priestley reportedly complained to producer Aaron Spelling about Shannen Doherty showing up late to tapings, among other things. She was knocked off the show in 1994.

• Actress Doherty married Ashley Hamilton, son of actor George Hamilton, in 1993, when she was twenty-two and he was eighteen. The two had only known each other a few weeks. Even Madonna was shocked, commenting about Doherty "marrying a teenager!" The couple divorced a short time later.

• Luke Perry, who plays Dylan, asked Los Angeles authorities in 1993 for permission to keep three pot-bellied pigs as pets in his L.A. home, which requires a special permit.

• Jennifer Grant, daughter of actors Cary Grant and Dyan Cannon, made her TV debut on the show as Celeste, a dating show contestant who goes out with Steve (Ian Ziering).

• Jennie Garth, who plays Kelly, married Dan Clark, drummer of the rock group Hoodwinks, in 1994.

• *Beverly Hills 90210* is produced by Darren Star, who also does the successful spin-off, *Melrose Place.*

• Andrea (Gabrielle Carteris) got pregnant in the 1993–1994 season, and actress Carteris was pregnant in real life too. She gave birth to her daughter, Kelsey, on May 11, 1994, the same night Andrea gave birth on the show.

• Joining the cast in 1994 was Tiffani-Amber Thiessen, as bad girl Valerie. Thiessen, who earlier played Kelly on *Saved by the Bell,* already knew many of the players on *Beverly Hills 90210,* as the real-life girlfriend of Brian Austin Green (David).

## DR. QUINN, MEDICINE WOMAN
**(January 1993– , CBS)**

Queen of the TV miniseries Jane Seymour, she of long-haired, British-accented gorgeousness, plays a frontier doctor, Michaela Quinn, in this family western medical drama. Dr. Quinn is from very proper Boston, and moves to Colorado Springs, Colorado, in the 1860s, to set up her practice, much to the chagrin of some residents who do not care for the fact the new doctor in town is female (there is a feminist message here).

Once entrenched, Dr. Quinn quickly becomes the foster mother of three orphans and starts a romance, of sorts (Dr. Quinn is a virgin), with hunky mountain man Byron Sully (Joe Lando), whose other favorite companion is a wolf. (The two were wed in a May 1995 episode.)

Critics generally dislike the show—*People* magazine called it "more dismal escapist fare from CBS." But by its second year, the show was on the air in seventy-five countries. It was CBS's biggest hit in its Saturday night spot since the 1970s.

*Dr. Quinn, Medicine Woman* is shot on location, not in Colorado, but at the Paramount Ranch in Agoura, California. The show uses an old western town set-up, a relic of a 1950s movie production, and other locations at the movie ranch, which is operated by the National Park Service.

## What to See in Agoura, California

▶ **Paramount Ranch,** Cornell Road (Kanan exit off the Ventura Freeway). Open daily to the public, 8 A.M. to sunset, this park is the only property of the National Park Service that is operated as a movie ranch. The land was previously owned by Paramount Studios (the National Park Service acquired the ranch in 1981). Visitors can see the park's old western town movie set, and may catch a glimpse of *Dr. Quinn, Medicine Woman* in production. The show does location work at the ranch on weekdays. Movies have been filmed on the property since the 1920s, and once a month, park rangers offer tours detailing the various production locations. Other shows that have filmed here include *Bat Masterson, The Rifleman, The A-Team, MacGyver,* and *The Six Million Dollar Man.* Call 818-597-9192.

## Fun Facts

• Jane Seymour gained fame in miniseries including *War and Remembrance* and *East of Eden.* She won an Emmy for her

*Dr. Quinn, Medicine Woman* is shot on location in an area operated by the National Park Service. *(Photo by Jean Bray, courtesy of the National Park Service.)*

supporting role in *Onassis: The Richest Man in the World,* in 1988.

• According to *TV Guide,* actress Seymour is a real nice lady, who invited cast and crew of *Dr. Quinn, Medicine Woman* to stay at her Malibu guest house in the aftermath of the L.A. earthquake of 1994.

• Seymour has been married four times, and her current husband is James Keach, brother of actor Stacey Keach. James Keach directed some episodes of *Dr. Quinn, Medicine Woman.*

• It's also no secret that Jane Seymour and her hunky costar Joe Lando, who plays Sully, briefly dated after the pilot episode of *Dr. Quinn, Medicine Woman* was filmed.

• Jane Seymour published a book in 1983, called, *Jane Seymour's Guide to Romantic Living.*

• Joe Lando previously shone his blue eyes and fabulous smile on the soap opera *One Life to Live.*

• Before actress Seymour was cast as Dr. Quinn, others considered for the part reportedly included Mel Harris *(thirtysomething).*

## EMERGENCY!
**(January 1972–September 1977, NBC)**

Jack Webb *(Dragnet)* produced this realistic-looking show, depicting the operations of a paramedic unit attached to the Los Angeles County Fire Department. Paramedics Roy DeSoto (Kevin Tighe) and John Gage (Randolph Mantooth) conducted death-defying rescue operations on the show, helping people stranded in cars in the Pacific Ocean, thwarting suicide attempts, and saving window washers trapped on top of skyscrapers. The show was filmed in cooperation with the L.A. Fire Department, whose own innovative paramedic unit has since been duplicated around the country.

*Emergency!* was praised by teachers, politicians, and public affairs groups for making communities aware of the benefit of paramedic activities. It was particularly popular with younger viewers. Stories on the show were fictional, but based on case studies and real incidents. And people actually learned lifesaving techniques from watching the program. For instance, a boy whose clothes caught on fire said he rolled on the ground, having witnessed a similar reaction on the show, and was spared from severe burns. And in 1972, the Tucson, Arizona Fire Department added a procedure learned from viewing the show, which involved using helmet straps to hold face masks in place.

The show filmed shots at a real L.A. County fire station, at a real L.A. County hospital, and sometimes even at the real L.A. County Fire Headquarters, as well as at other locations around L.A. and on a set at Universal Studies. *Emergency!* also sometimes left the L.A. area for location shoots, including an episode filmed in Seattle.

## WHAT TO SEE IN CARSON, CALIFORNIA

▶ **L.A. County Fire Station 127,** 2049 223rd Street (off Wilmington, and across from the Arco Refinery). The outside of this real station was seen as Station 51 on the show. Visitors are invited to take pictures outside, and if the gang inside isn't busy you

L.A. County Fire Station 127. *(Photo courtesy of the Los Angeles County Fire Department.)*

might be invited inside, as well. The inside appeared occasionally on the show, but most interiors were done on a set.

## What to See in Torrance, California

▶ **Harbor UCLA Medical Center,** 1000 West Carson Street. This real-life hospital appeared as Rampart General Hospital on the show.

## What to See in Los Angeles, California

▶ **L.A. County Fire Headquarters,** 1320 Northeastern Avenue (in East L.A.). This building, which houses the administrative offices of the county fire department, also sometimes appeared on the show.

## Fun Facts

- Rampart General was the name of the hospital on *Emergency!* Rampart was also the name of the police division on *Adam 12*, another Jack Webb production.
- Bobby Troup, who played Dr. Joe Early, and Julie London, who played nurse Dixie McCall, were married in real life. She was earlier married to producer Jack Webb.
- Bobby Troup is also a musician, whose hits have included "Route 66," "Girl Talk," and "Daddy."
- Robert Fuller, who played Dr. Kelly Brackett, was earlier on *Laramie* and *Wagon Train*.
- Randolph Mantooth, who played John Gage, is the son of a full-blooded Seminole Indian. He later appeared on the soap *Loving* as Alex Masters.
- Olympic gold medalist Mark Spitz, actress Linda Gray *(Dallas)*, and actor Bobby Sherman *(Here Come the Brides)* were among the show's guest stars.
- A big fan of the show was Senator Alan Cranston (D-Calif.).
- A cartoon version of the show, *Emergency + 4*, aired from 1973 to 1976, on NBC.

## FANTASY ISLAND
**(January 1978–August 1984, ABC)**

Like *Love Boat*, which it followed on Saturday nights, *Fantasy Island* provided a place where visitors' dreams, many romantic (Aaron Spelling was coproducer, after all), could come true. In this case, the place was a small tropical island, not a luxury cruise ship; a setting that allowed for a sense of adventure in some of the fantasies.

Movie actor Ricardo Montalban played Mr. Roarke, the mysterious owner of the island, whose power sometimes seemed magical.

He was assisted by Tattoo (Herve Villechaize) who shouted in a French accent, "The Plane! The Plane!" and rang a bell whenever a new batch of guests arrived. Guest stars appeared as the island's visitors.

The show filmed some episodes on location, not on an island, but at the Los Angeles State and County Arboretum in Arcadia. And Tattoo rang the bell in the tower of a real-life Queen Anne Victorian cottage at the botanical gardens. The cottage was later recreated on a sound stage. Interiors for the show were also done at the studio.

The arboretum's lake also appeared on the show. It was the landing site of the pontoon plane that brought guests to the island.

### WHAT TO SEE IN ARCADIA, CALIFORNIA

▶ **Los Angeles State and County Arboretum,** 301 N. Baldwin Avenue. The Queen Anne Victorian cottage seen on the show was built in 1886, as a guest house on the estate of Elias Jackson Baldwin, otherwise known as "Lucky" Baldwin. The interior of the cottage is not open to the public, but can be seen through viewing windows. The park also has a 127-acre botanical garden and an indoor tropical greenhouse. The arboretum is open from 9 A.M. to 4:30 P.M., seven days a week. A fee is charged. Call 213-681-8411.

The Queen Anne Victorian cottage at the Arboretum of Los Angeles County. *(Photo by Eileen Ames, Arboretum of Los Angeles County.)*

## Fun Facts

- Ricardo Montalban, who is from Mexico, also played the leading man in many films. But to many, Montalban's most memorable movie role was that of Khan in *Star Trek: The Wrath of Khan* in 1982.
- On TV, Montalban also appeared in *How the West Was Won*, for which he won a best supporting actor Emmy in 1978, and on *The Colbys*.
- Herve Villechaize, who suffered from ailments for years, died at age fifty, in 1993, the victim of a self-inflicted gunshot wound. In real life, the 3-foot, 11-inch, Paris-born actor was known as a bit of a ladies' man.
- Before his *Fantasy Island* role, Villechaize appeared in the James Bond movie *The Man with the Golden Gun*.

## I LOVE LUCY
**(October 1951–September 1961, CBS)**

"Wahhhhhhhh!"

CBS originally wanted this comedy shot live in New York. But Desi Arnaz wanted the show shot in L.A. The network also did not particularly want Arnaz, who was Cuban, to play the husband of all-American Lucille Ball, who happened to be his real-life wife. The network lost on both points.

*I Love Lucy* was filmed in Hollywood, before a live audience. And if anything, Arnaz ended up exaggerating his Cuban accent to play off wacky red-haired Ball.

On the show Lucy Ricardo (Ball), a housewife, and Ricky Ricardo (Arnaz), a nightclub performer, lived at 623 East 68th Street in New York. The address was fictional. Permanent sets created in Hollywood, for the show, included the apartment of the Ricardos and The Tropicana Club, where Ricky performed.

According to Bart Andrews, in *The "I Love Lucy" Book*, the show also sometimes shot footage on location. The first time was in 1955, when the Ricardos and their landlords, Ethel (Vivian Vance) and Fred (William Frawley), were seen traveling cross-country in a convertible, on their way to Hollywood (the setting for several episodes). Shots were filmed of the actors on the road in a new Pontiac convertible (General Motors paid a promotional fee, and chipped in several extra cars, for the plug).

Around the same time, in real life, the Arnaz family moved into a new house. And their own stately Williamsburg-style home appeared in an episode called "The Tour," which was part of the Hollywood story line. In the episode Ethel and Lucy took a tour of stars' homes in Beverly Hills. At Richard Widmark's house, Lucy decided to scale a wall to try to get a souvenir grapefruit from a tree on the other side. Of course, she got caught. The white wall of the Arnaz's real house was duplicated on the set, and shots of the house itself were also seen in the episode.

Lucy, Ricky, Ethel, and Fred traveled together again on the show, in 1956, this time to Europe. But that trip was not shot on location. A special promotional fee paid by America Export Lines to promote its ship, the *Constitution*, paid for construction of a special set for the cruise episodes, according to writer Andrews. The *Constitution*'s deck, stateroom, and bar were duplicated on the set.

In one of the Europe-based episodes, Lucy, in Italy, thought a director wanted her for the part of a wine grape stomper (he really wanted her to play an American tourist), so she decided to impress him with her wine-stomping ability. A real wine stomper was brought south to Hollywood from a winery in Napa Valley, California, to add authenticity to the scene.

*I Love Lucy* went on location in 1958, to Las Vegas, and also filmed all or part of later episodes in locations including Sun Valley, Idaho, and Lake Arrowhead, California (although the latter appeared as Alaska on the show).

### What to See in Beverly Hills, California

▶ **Lucy and Desi's real house,** 1000 Roxbury Drive (at Lexington Avenue). Lucy fell in love with this white, Williamsburg-style house, according to *Lucy in the Afternoon* author Jim Brochu, because it reminded her of those in her native Jamestown, New

Lucille Ball's former house in Beverly Hills. *(Photo by Fran Golden.)*

York. The Arnaz's neighbors in this choice neighborhood included Jack Benny *(The Jack Benny Show)*, who lived at 1002 Roxbury Drive.

## Fun Facts

- CBS originally wanted Richard Denning *(Hawaii Five-O)*, who had costarred with Lucille Ball in the 1940s radio show, *My Favorite Husband*, on which *I Love Lucy* was based, to play her husband on the show.
- Character Lucy's maiden name is MacGillicuddy, and like Lucille Ball, she is from Jamestown, New York.
- During its first six seasons, *I Love Lucy* was such a hit it was always rated as one of the top three most popular shows.
- Ethel (Vivian Vance) had three different middle names on the show, Mae, Louise, and Roberta.
- It was reportedly in actress Vance's contract that she would have to stay overweight and wear dumpy dresses.
- William Frawley, who played grumpy Fred, was known to be a heavy drinker.
- The show was produced by Desilu, a company owned by the Arnazes, which also produced such hit shows as *Make Room for Daddy* and later *Star Trek*.
- Lucille Ball was accused in the early 1950s of being a Communist. The actress explained that years before she had filled out a voter registration form as a Communist, but she said she did so only to make her grandfather happy.
- Lucille Ball and Desi Arnaz met on the set of *Too Many Girls*, a movie version of the Rodgers and Hart musical. On the show, Lucy and Ricky met on a blind date.
- According to writer Bart Andrews, the live audience of *I Love Lucy* laughed the longest in the episode called "Lucy Does the Tango," in 1957, as the result of a routine that involved Lucy and raw eggs.
- Lucille Ball was five months pregnant with Lucie Desiree Arnaz when the pilot for *I Love Lucy* was filmed.
- The first issue of *TV Guide*, which began publication in April 1953, bore a picture of the Arnaz's second child, newborn Desi Arnaz, Jr., on its cover.
- Early in her career, Ball once went by the name of Diane Belmont, a name she derived from the Belmont Race Track.

## KNOTS LANDING
**(December 1979–May 1993, CBS)**

Just as addictive as *Dallas,* from which it was spun off, and *Dynasty, Knots Landing* gave nighttime soap fans middle-class characters and backyard barbecues as an alternative to the flowing champagne and life-styles of the rich and famous of other shows. That's not to say there weren't tantalizing story lines, however. There was plenty of sex, scandal, and secrets involving the residents of the Seaview Circle cul-de-sac.

The show was never quite as campy as its wealthier rivals, but it managed to outlast them on the air, running fourteen seasons.

The setting for *Knots Landing* was suburban Southern California, and in case you missed the point, most of the women on the cul-de-sac were blondes.

The initial episodes of the show had Gary Ewing (Ted Shackelford), the black sheep of the *Dallas* Ewing clan and father of Lucy Ewing (Charlene Tilton), moving to Knots Landing, California, after remarrying his ex-wife Val (Joan Van Ark). Their off-again, on-again relationship continued for the entire length of the series.

The star villain of *Knots Landing* for much of the show's run was Abby Cunningham (Donna Mills), sister of Sid Fairgate (Don Murray), owner of Knots Landing Motors. (Sid is married for a time to Karen, played by Michele Lee.) After she causes as much trouble as is humanly possible, and adds both Ewing and Sumner to her last name through marriage, Abby leaves the cul-de-sac (in 1989) to become a trade ambassador to Japan.

The show's plots got more and more absurd, but along the way dealt with environmental issues and homelessness, when Anne (Michelle Phillips) was without a home, along with kidnappings and murders.

In 1991, with ratings declining, and with several of the show's writers leaving to work on a new show, *Homefront,* the producers pulled the show off the air for a few weeks to try to create story lines that would bring it back to reality. When the show returned, Greg Sumner (William Devane) had an affair with Anne while flirting with her vixen daughter Paige (Nicollette Sheridan), who is the illegitimate daughter of Mack (Kevin Dobson).

Guests on the show included Gary's brother, J. R. Ewing (Larry Hagman), and Gary and Val sometimes visited the *Dallas* set, too, even though there were some hard feelings over the publishing of Val's book, *Capricorn Crude,* which was loosely based on the Ewing clan.

A lot of *Knots Landing* was filmed in a Hollywood studio, but private homes around L.A. were also used. Unfortunately, the owners want the addresses to remain private, according to location manager Jeffrey Smith. Fans of the show will, however, recognize Marina del Rey, the world's largest man-made small craft harbor, where harbor shots for the show were frequently filmed. Some other location shots were done in Oregon, Smith said.

## What to See in Marina del Rey, California

▶ **Burton Chace Park,** Mindanao Way. This park appeared as the marina at the Lotus Point luxury resort development on the show. The eight-acre park is at the end of one finger of Marina del Rey and offers docks for visiting boaters. The park was created on ground that was previously a mosquito-infested swamp, according to locals, and is named for a county supervisor who spearheaded the beautification project. Other shows that have featured the park include *Simon & Simon, Baywatch, Charlie's Angels,* and *Remington Steele.*

For visitor information on Marina del Rey, call 310-305-9545.

## Fun Facts

• Michelle Phillips, who played Anne Matheson, was one of the original Mamas and Papas. One of that group's hits, "This Is Dedicated to the One I Love," was heard in a love scene between Anne and Mack (Kevin Dobson).

• In addition to *Capricorn Crude,* writer Valene Ewing (Joan Van Ark) published *Nashville Junction.*

• Julie Harris, the Tony– and Emmy–award winning actress, played Valene's mother, Lilimae Clements. Movie actor Alec Baldwin appeared on the show (early in his career), playing Lilimae's son, and Val's half-brother, Joshua Rush.

The Burton Chase Park in Marina del Rey was known as Lotus Point on *Knots Landing*. *(Photo by Greg Wenger, Wenger International Photography.)*

- Actress Lisa Hartman Black (then just Lisa Hartman) played two parts, Ciji Dunne, a singer, and (after Ciji is murdered) look-alike Cathy Geary, a waitress.
- Mack's (Kevin Dobson) real name on the show is Marion Patrick MacKenzie.
- Movie star Ava Gardner guest-starred as Ruth Galveston, mother of Greg Sumner (William Devane).
- Gary (Ted Shackelford) is the father of Val's (Joan Van Ark) twins, Bobby (Christian and Joseph Cousins) and Betsey (Kathryn and Tiffany Lubran, and later Emily Ann Lloyd), but they weren't married when Val gave birth to them.
- Actress Nicollette Sheridan (Paige) was married briefly in real life to actor Harry Hamlin *(L.A. Law)*, and later fell for singer Michael Bolton. She was raised with the help of Telly Savalas *(Kojak)*, who had a long-term relationship with her mother, British actress Sally Adams.

## LIFE GOES ON
**(September 1989–August 1993, ABC)**

This critically acclaimed but low-rated show, featured in a lead role actor Chris Burke, who has Down's syndrome in real life, and who played mentally handicapped Corky on the show. On *Life Goes On,* Corky is mainstreamed at Marshall High School, the same school attended by his teenage sister, Becca (Kellie Martin). Both Corky and Becca deal with issues of growing up, including falling in love and sibling rivalry. Corky dates Andrea Friedman (Amanda Swanson), who also has Down's syndrome. Becca falls for Jesse (Chad Lowe) who is HIV-positive.

Like *The Wonder Years,* this heartfelt family drama sought realism by shooting some scenes in real L.A.-area high schools.

### What to See in Van Nuys, California

- **Van Nuys High School,** 6535 Cedros Avenue. This suburban high school is seen inside and out on both *Life Goes On* and *The Wonder Years*. The school was built in 1914, and has a student population of about 2,800. Some lucky students appeared as extras on the show. Famous alumni of the school include Robert Redford, Natalie Wood, Paula Abdul, Stacey Keach, and Jane Russell. Marilyn Monroe attended the school too, but did not graduate.

- **Grant High School,** 13000 Oxnard Street. Indoor and outdoor scenes for *Life Goes On* and other high school shows, including *The Wonder Years, Parker Lewis Can't Lose,* and *Beverly Hills 90210,* show scenes of this large suburban high school of more than 3,000 students. Famous alumni include Tom Selleck *(Magnum, P.I.).*

### Fun Facts

• Patti LuPone, who played Libby Thatcher, starred in *Evita* on Broadway and had the lead in the London production of Andrew Lloyd Webber's *Sunset Boulevard.* She was replaced by Glenn Close for the show's U.S. run. Since her character, Libby, is a former singer on the show, LuPone also sang occasionally on *Life Goes On.*

• Chad Lowe, who played Jesse, is the younger brother of movie actor Rob Lowe. He earlier appeared on TV in the sitcom *Spencer*, but quit that show after only six episodes due to creative differences.

• Tim Wood, the manager of actor Lowe, died of complications from AIDS, and Lowe told *People* magazine he drew inspiration for the part of Jesse from Wood's experiences.

• Andrea Friedman, the actress with Down's syndrome who played Amanda, studied psychology at Santa Monica College.

• The theme song for the show is "Ob-La-Di, Ob-La-Da," by John Lennon and Paul McCartney.

• Kellie Martin, who played Rebecca Thatcher, later continued her acting career while also attending Yale.

• Chad Lowe won an Emmy for his role as Jesse in 1983.

## M*A*S*H
**(September 1972–September 1983, CBS)**

An Army corporal peers into the mountains as a chopper bearing war's wounded prepares to land. The theme song, "Suicide Is Painless," plays in the background. Welcome to the front lines of Korea, where the United States defended South Korea against Chinese-backed North Korea from 1950 to 1953, or at least the dark comedy TV version thereof.

Inside the show's army-green tents, beneath high-intensity lights and often with the sounds of bombs going off in the background, the doctors of the show's 4077th Mobile Army Surgical Hospital (M.A.S.H.) treat casualties of the conflict. And their work is grueling and gruesome. But they manage to get by with the help of humor, moonshine, cruel pranks, and an ever-present supply of sexy nurses.

Inspired by the characters of the hit Robert Altman film of the same name (which stars Donald Sutherland and Elliott Gould), which was in turn inspired by a book by a real M.A.S.H. unit doctor, the TV show's scripts mentioned actual details of the Korean conflict (managing to stretch the scenario for eleven years). But *M*A*S*H* is as much about the lives, loves, and feelings of the characters as their war experiences.

Alan Alda, previously a New York stage actor, played the TV

version of Benjamin Franklin "Hawkeye" Pierce, the lead role, as an irreverent and often smirking, yet brilliant and empathetic, surgeon, joking his way through the dark aspects of the war. Pierce refers to the war as "a costume party." And having been drafted, he is not a very willing participant in the battle, putting him and his cronies in continuous comedic conflict with the commissioned officers of the unit.

The Vietnam War was going on when *M*A*S*H* debuted, and Pierce's antiwar jokes and comments were as pertinent to Vietnam as to the conflict waged twenty years earlier.

The show was not an immediate hit, and was almost canceled after its first season. But after it caught on, *M*A*S*H* was in the list of top ten shows for nine out of its eleven years. The 1983 finale of *M*A*S*H,* in which Hawkeye recovered from a nervous breakdown as the war ended, broke records for audience size, estimated at 125 million viewers.

In recognition of the show's popularity, two complete stage sets, "The Swamp" and the operating room, were accepted as donations by the Smithsonian in Washington. Included in the museum's collection are Hawkeye's martini glass and the tabletop moonshine still, as well as Winchester's (David Ogden Stiers) record player and costumes from the show.

Outdoor Korea scenes on *M*A*S*H* were filmed not in the Asian nation, but at a movie ranch owned by Twentieth Century-Fox and located in the Santa Monica Mountains. The ranch is now a state park, and open to the public.

### What to See in Calabasas, California

▶ **Malibu Creek State Park,** 1925 Las Virgenes Road (off the Ventura Freeway). This state park was formerly the Century Ranch, part of Twentieth Century-Fox. The area used for filming Korea scenes on *M*A*S*H* is about a 2½-mile hike from the main gate, in a canyon area. Two rusted-out jeeps used on the show remain at the site, and a marker notes the area's historic value. The park also has a visitors center, which has some memorabilia on display from the show, along with natural history exhibits. Dog tags that say "I visited the *M*A*S*H* site," are for sale. The visitors center is open from noon to 4 P.M. on weekends. The park itself is open daily, 8 A.M. to sunset, and there is a campground for overnight stays. There is a fee for parking. Call 818-880-0367.

## Fun Facts

• The book *M*A*S*H* was written by a doctor using the pseudonym Richard Hooker, who had served in Korea. Like the main character, Benjamin Franklin "Hawkeye" Pierce (Alan Alda), the doctor was from tiny Crabapple Cove, Maine.

• Two actors, Gary Burghoff, who played Radar, and Tim Brown, who played Spearchucker Jones, appeared in the movie and on the TV show.

• Radar (Burghoff) always heard the hum of the helicopters before anyone else.

• As his star rose, Alan Alda, who played Hawkeye, also wrote and directed several episodes of the show, winning Emmy awards for acting, directing, and writing,

• Actress Joan Van Ark *(Knots Landing)* played Lieutenant Erika Johnson in the show's second season.

• Popular products sold bearing the *M*A*S*H* logo included T-shirts and dog tags.

• The show had a less successful sequel, *AfterMASH*, which starred Colonel Potter (Harry Morgan), Klinger (Jamie Farr), and Father Mulcahy (William Christopher) as civilians working at a V.A. hospital in Missouri.

• The show *Trapper John, M.D.* picked up a character featured on *M*A*S*H*, but starred Pernell Roberts *(Bonanza)*, not Wayne Rogers, who played the part in *M*A*S*H*.

• The United States lost 142,091 men in the Korean conflict.

• Alan Alda is the son of actor Robert Alda, whose career included hosting several 1950s game shows and appearing on the soap opera *Love of Life*. He also appeared with his son on *M*A*S*H*.

• During the final episode of *M*A*S*H*, a real fire that swept through the show's set was written into the script, which also made reference to a brewing conflict in a country called Vietnam.

## THE ROCKFORD FILES
**(September 1974–July 1980, NBC)**

James Garner, who had earlier starred in the satirical 1950s western *Maverick*, starred in *The Rockford Files* as Jim Rockford, a pardoned ex-convict (he was never really guilty of any crime), turned private eye.

Dry-humored Rockford lived in a trailer in Malibu and generally worked with clients whose cases had already been closed by the police. His services didn't come cheaply, with Rockford charging $200 a day, plus expenses.

The private eye didn't like guns, but kept one handy in his cookie jar, just in case.

He was helped out by his father, Rocky (Noah Beery), and by his lawyer girlfriend, Beth (Gretchen Corbett). Rockford's former cell mate, Angel Martin (Stuart Margolin), was usually nearby causing trouble.

*The Rockford Files* was an Emmy-award-winning show, but went off the air suddenly in 1980 after Garner, who was experiencing health problems (mostly with his back and knees), quit. Lawsuits ensued, with Universal Television later settling with Garner out of court.

The dispute patched up, the studio in 1994 helped produce the resurrection of *The Rockford Files*, in the form of six new TV movies.

## What to See in Malibu, California

- **Sand Castle Restaurant,** 28128 Pacific Coast Highway. The parking lot next to this casual restaurant on the beach is where Jim Rockford parked his trailer on the show. The TV movies also did some filming at the site. The restaurant serves breakfast, brunch, lunch, and dinner daily. Seafood is a specialty. Open 6 A.M. to 10 P.M. daily (the bar is open to 11 P.M.), and until 11 P.M. on Saturday (the bar is open Saturday until midnight). Call 310-457-9793.

## Fun Facts

- Early in his career, James Garner, whose real name is James Bumgarner, reportedly modeled bathing suits.
- Stuart Margolin won two Emmys (in 1979 and 1980) as best supporting actor for his role as sleazy ex-con Angel Martin.
- Both Margolin and Joe Santos (Detective Dennis Becker) were reunited with Garner in the Rockford TV movies.

• Noah Beery, who played Rockford's father, Rocky, was suffering the lingering effects of a stroke and was too ill to appear in the first of the TV movies, *The Rockford Files: I Still Love L.A.*, in 1994. Beery died shortly before the movie aired, in November 1994, and the project was dedicated to his memory.

• Actor Beery's earlier appearances included *Circus Boy,* the 1950s show starring a young Mickey Dolenz of *The Monkees*.

• Tom Selleck *(Magnum, P.I.)* appeared on *The Rockford Files* as Lance White, a competing, all-too-perfect private eye.

• James Garner won an Emmy for his role as Jim Rockford in 1977.

• The theme song from the show, "The Rockford Files," by Mike Post and Peter Carpenter, was a hit song in 1975.

• Actor Garner's movie roles include *Murphy's Romance,* for which he was nominated for an Oscar, *Victor/Victoria,* and the movie version of *Maverick* (although Mel Gibson played the title role of Bret Maverick in the 1994 film).

• Not all Garner's TV appearances were in hit shows. His 1970s western, *Nichols*, lasted only one season on NBC.

## THIRTYSOMETHING
**(September 1987–September 1991, ABC)**

The *Daily News* called this dramatic series "a life-style." The *L.A. Times* complained it was "yuppy angst." Some fans loved the seven main characters, three single and four married, all white and all in one way or another questioning their lives. Others couldn't stand them, but watched the show anyway.

The show was written, directed and acted by thirtysomethings, and many of the actors and crew were friends in real life. A lot of the ideas for the show, such as career versus family debates, came from the real lives of the cast and crew.

Hope (Mel Harris) and Michael (Ken Olin) were never married in real life, but Ken Olin was married to Patricia Wettig, who played Nancy, wife of Elliot (Timothy Busfield) on the show. Olin, who directed some episodes of the show, and Wettig met producer Marshall Herskovitz through their kids, who went to preschool together.

In California, therapists felt the show so reflected problems of the babyboomer generation that they used episodes of *thirtysomething* with patients. One recommended having new parents view the show as a means of getting them to talk about their feelings, reported *Psychology Today.*

Herskovitz, who produced the show with Edward Zwick (the two had earlier written for *Family*), told the magazine the use of the show for therapy made sense, as *thirtysomething* was about "inner conflicts."

But not everyone liked the on-screen yuppy analysis.

Comedian Jay Leno said of the show, "First I see the wife, and she's whining, 'What about my needs?' Then they cut to the husband, and he's whining, 'What about my needs?' And I'm sitting here saying, 'What about my needs? I want to be entertained. Can't you blow up a car or something?'"

The show could be funny. It could also be depressing, which is another reason it attracted attention from psychologists. Gary's (Peter Horton) sudden death in a car crash, coming right after Nancy (Wettig) received a good prognosis in her battle with ovarian cancer, was so traumatic to some viewers they sought psychological help.

Actress Polly Draper, who played Ellyn, told a reporter the cast watched the episode in which Gary died together, at actress Melanie Mayron's (she played Melissa) house, and they all cried. So did a lot of viewers.

While the show was often sentimental, it was not shy of controversy either. In a 1989 episode, two gay characters were shown after sex, talking about AIDS. The conservative Mississippi-based American Family Association called for an advertisers boycott, and ABC reportedly lost about $1 million in advertising. But the producers were not put off by the protest. The characters returned as guests at a New Year's Eve party in an episode a year later.

The show was not filmed in Philadelphia, where the fictional characters were based, but mostly in Studio City, California. And the house seen as Hope and Michael's is really on a tree-lined street in South Pasadena.

The *thirtysomething* house in South Pasadena. *(Photo by Fran Golden.)*

## What to See in South Pasadena, California

▶ **The house at 1710 Bushnell Avenue.** Owner Dennis Potts says a location scout knocked on the door of his Craftsman bungalow-style house one day and asked if it could be used on the show. The house became Hope and Michael's, and the subsequent filming inside and outside helped Potts pay the college tuition of one of his kids. The house was built in 1902, and is similar inside to that seen on the show except for an added front staircase in the fictional version. In the show's second year, the interior of the Potts house was re-created on a Studio City set. Potts said when the crew shot exteriors they wrapped trees in his yard to make them look more like Philadelphia varieties. A palm tree was made to look like bushes, for instance. The block the house is on has also been seen in several movies. The house across the street at 1711, for instance, was featured in *Back to the Future*.

## Fun Facts

• Peter Horton, who played Gary, was once married to Michelle Pfeiffer and went into directing after leaving the show.

• Geffen Records released *The Soundtrack from thirtysomething* in 1991. It had eighteen cuts including Ray Charles's 1959 version of "Come Rain or Come Shine," which had been played at Ellyn's (Polly Draper) wedding.

• Actress Draper went to Yale. In real life, her main squeeze was Michael Wolfe, Arsenio Hall's music director on *The Arsenio Hall Show.*

• "Thirtysomething," the clothing label, was introduced by Apparel Resources International. The women's collection was priced from $25 to $325 and the men's collection from $25 to $600. Offerings included jeans, vintage-looking jackets, denim shirts, and retro-style ties.

• A New York public relations firm looking for account executives got a lot of applicants for a job by posing in a newspaper ad as D.A.A. (Drentell Ashly Arthur), the name of the ad firm on the series.

• The character of Miles Drentell (David Clennon), the boss of the ad firm on the show, is a takeoff on William Drenttel, a real ad man and a partner in the New York agency Drenttel Doyle Partners, who is also an old friend of producer Edward Zwick.

• The production company formed by Zwick and Herskovitz that produced the show is called Bedford Falls, named after the town in the Jimmy Stewart movie *It's a Wonderful Life.*

## THE WONDER YEARS
**(March 1988–May 1993, ABC)**

Fred Savage starred as young Kevin Arnold in this nostalgic comedy-drama about growing up in the late 1960s and early 1970s. Innocent-looking (but not always acting) Kevin listens to the Beatles and follows events in the Vietnam War, while dealing with dating, his siblings (a flower child and a bully), a brutish athletic coach, and other suburban teenage issues.

Actor Savage was twelve when the show began and literally

grew up before viewers' eyes, gaining a foot in height and developing a deeper voice. Also growing up on the show was Danica McKellar as Winnie Cooper, Kevin's friend and sometimes love interest.

The show created period realism by using real news clips of sixties and seventies events and popular music from the time. *The Wonder Years* also achieved a real look by filming at several actual Los Angeles-area high schools.

### What to See in Burbank, California

▶ **Burroughs High School,** 1920 Clark Avenue. *The Wonder Years* series shot scenes here until 1991, when the Burbank Unified School District pulled the plug because it deemed the filming too disruptive during school hours, according to the *Los Angeles Times*. The school principal said he was surprised because he thought the relationship was good. Famous alumni of this high school include Ron Howard *(Happy Days)*, Debbie Reynolds, and Tim Matheson.

### What to See in Van Nuys, California

▶ **Van Nuys High School,** 6535 Cedros Avenue. This high school, in a suburb north of L.A., was used as a location for both *The Wonder Years* and *Life Goes On*.

▶ **Grant High School,** 13000 Oxnard Street. This high school was used on *The Wonder Years* and also appeared on *Life Goes On*, *Beverly Hills 90210*, and *Parker Lewis Can't Lose*.

### Fun Facts

• Actor Daniel Stern narrated *The Wonder Years*, as the voice of Kevin as an adult. He is best known for his role as a crook (with Joe Pesci as his partner) in the *Home Alone* movies. Stern earlier appeared in the movie *Breaking Away*.

• Actor Robert Picardo, who appeared on *The Wonder Years* as fearsome Coach Cutlip, was, at the same time, also appearing as Dr. Dick Richard on *China Beach*.

• Winnie's (Danica McKellar) real name is Gwendolyn.

• A former wardrobe assistant on the show, Monica Long, thirty-two at the time, filed a sexual harassment suit against Savage and costar Jason Hervey, who played Kevin's big brother, Wayne. Representatives for the actors responded by calling Long a disgruntled employee.

• Fred Savage is from Glencoe, Illinois, although his family moved with him to California when his career took off. He got his break in show business in a Pac-Man Vitamin commercial. Savage was reportedly paid $75,000 an episode by the end of *The Wonder Years.*

• There's also another Savage on TV. Fred Savage's younger brother, Ben, who stars in ABC's *Boy Meets World.*

## FYI: ALSO IN LOS ANGELES COUNTY AND ENVIRONS

*Benson* (September 1979 to August 1986, ABC), starring Robert Guillaume as Benson, a butler turned politician, takes place in an unnamed state. But the estate shown as the governor's mansion on the show is really located in Pasadena.

### What to See in Pasadena, California

▶ **The Mansion** at 1365 S. Oakland Avenue. This hillside estate is seen as the governor's mansion on *Benson.*

On *Knight Rider* (September 1982 to August 1986, NBC), crime fighter Michael Knight, played by David Hasselhoff *(Baywatch)*, works out of the Knight Industries' Foundation for Law and Government, based in an impressive California mansion.

The real palatial estate shown on the show is in Pasadena, and has also appeared in more than five hundred movies, including the Marx Brothers' *Duck Soup,* as well as in numerous TV shows.

### What to See in Pasadena

▶ **Morton Mansion/Arden Villa,** 1145 Arden Road (near the Cal-Tech campus). In addition to serving as the headquarters on

This Pasadena mansion was seen on *Benson*. *(Photo by Fran Golden.)*

*Knight Rider,* this mansion, built in 1915, was the scene for the famous fight between Krystle (Linda Evans) and Alexis (Joan Collins) on *Dynasty,* in which Krystle pushed Alexis into a fish pond. The 20,000-square-foot mansion also appeared on *Hart to Hart* and *Flamingo Road.* The interior has also been used numerous times to double as the White House. The mansion is difficult to see from the road and is not open to visitors.

The sitcom *Step by Step* (September 1991 to , ABC) takes place in Port Washington, Wisconsin, and is filmed in a studio. But the fabulous roller coaster seen in the intro of the show, which brought back stars Suzanne Somers *(Three's Company)* and Patrick Duffy *(Dallas)* to the weekly TV show lineup, is really in sunny southern California.

The Colossus roller coaster at Six Flags Magic Mountain. *(Photo courtesy of Six Flags Magic Mountain.)*

## What to See in Valencia, California

- **Six Flags Magic Mountain,** 26101 Magic Mountain Parkway (off the I-5 freeway). The roller coaster seen on the show is the Colossus. It's just one of many rides at this amusement park, which is open daily, 10 A.M. to sundown, from late March to mid-September. Admission is charged. Call 805-255-4100.

# ELSEWHERE IN CALIFORNIA

As natives know and visitors discover, California is not all Los Angeles and bikini beaches. There are deserts and mountains and rural western settings only hours outside of La La Land, and they have naturally been tapped to appear as backdrops on TV shows.

Death Valley, the desert and canyon location that is the hottest place in the United States, was the real-life setting seen on some episodes of *Death Valley Days*, which traces the history of the region.

And *Death Valley Days* was also among numerous westerns, including *Bonanza*, that featured, in some episodes, the Alabama Hills, an area of odd moonlike rock formations near Lone Pine, California, and the High Sierras. The landscape was also seen in at least one episode of *The Twilight Zone*.

Another area of interesting rock formations, the Vasquez Rock County Park (off Route 14 in Agua Dulce, California) appeared as the landscape of other worlds on *Star Trek*.

Up north near Sacramento is another area that has attracted many TV production crews, Calaveras County. The Gold Rush area's historic western buildings have been a setting for TV shows including *Seven Brides for Seven Brothers* and *Highway to Heaven*.

In nearby Jamestown, California, Railtown, a state park where historic steam and diesel trains are displayed, was the setting seen in train scenes on *Petticoat Junction* and *Green Acres*. And visitors can view the real train that doubled for fictional Hooterville's *Cannonball* on the rural comedies.

Down south near San Diego, TV fans can visit the actual

Marine Corps base that was seen as fictional Camp Henderson on *Gomer Pyle, U.S.M.C.*

## DEATH VALLEY DAYS
**(1952–1970s, syndicated)**

This western anthology show featured an "Old Ranger" host and colorful stories of early western times, all based on fact, as gleamed from history books, as well as old newspaper articles and first-hand research. The main setting was Death Valley, the lowest and hottest place in the United States which stretches about 140 miles through eastern California into Nevada.

During the early years of the long-running show, many episodes were filmed on location in Death Valley, California, with filming also taking place in Studio City. Later, the stories moved out of Death Valley, and other western locations were used, including sites in Arizona and Utah. *Death Valley Days* also used the unusual rocky landscape of the Alabama Hills in nearby Lone Pine, California, for location footage.

Besides its western tales, the show is probably most noted for one of its hosts, actor Ronald Reagan. The pre-presidential Reagan actually was on the show only from 1965 to 1966, leaving for politics and his eventual election as governor of California. Reagan got the job thanks in part to his brother, Neil, a New York ad man, who worked for the McCann-Erickson agency. The ad agency handled the account of the Borax Company, which sponsored the show.

*Death Valley Days* was, in fact, created by McCann-Erikson, originally for radio, to promote Twenty-Mule Team Borax soap flakes. Borax is mined in the Death Valley region.

The first host of the TV show was Stanley Andrews, who held down the job for thirteen years (from 1952 to 1965). Other hosts included Robert Taylor, Dale Robertson, and Merle Haggard. The cast of the show changed weekly.

Early *Death Valley Days* episodes were written by a proper Rye, New York, woman, Ruth Cornwall Woodman, who worked for McCann-Erikson, and who also wrote the series for radio. Woodman, a Vassar graduate, researched the show's stories firsthand in the desert, becoming one of the leading authorities on Death Valley folklore. But she was not credited for many years as a writer on the show, because the producers worried that people would not buy the Old West stories if they knew an East Coast mother of two, married to a New York businessman, was behind the telling of the tales. Woodman, who was also an enthusiastic world traveler, retired in 1959, and died in 1970, at age seventy-five.

## WHAT TO SEE IN DEATH VALLEY, CALIFORNIA

▶ **Death Valley National Monument Visitors Center** (off Highway 190, Furnace Creek). The name of the valley originated in 1849, when a gold miner referred to the massive hot valley as Death Valley. The valley offers visitors views of canyons, mountains, and odd formations of rock. This visitors' center is a good place to start a tour. It offers natural history exhibits, including a display on borax mining, and information on the region. The

On location with *Death Valley Days,* and actor James Craig. *(Photo courtesy of the Ruth Cornwall Woodman Collection.)*

center is run by the National Park Service. It is open daily, 8 A.M. to 6 P.M. Call 619-786-2331.

▶ **Scotty's Castle** (on Route 5, Grapevine Canyon, about fifty-five miles north of Furnace Creek). This ranch house, also known as "Death Valley Ranch," is in the desert and was one of many real sites mentioned in *Death Valley Days* scripts. It was built by Chicago businessman Albert Johnson, but is named for his flamboyant friend, Scotty, whose real name was Walter Scott. Scotty was a former bronco rider for Buffalo Bill's Wild West Show, turned prospector and storyteller (among his stories was that the castle was his). Maintained by the National Park Service as a National Historic Landmark, the castle, which was never quite finished (construction halted in 1931), is operated as a living history museum, complete with costumed characters. Tours are offered daily, 9 A.M. to 5 P.M. A fee is charged. Call 619-786-2392.

For information on other sites in Death Valley, call the Death Valley Chamber of Commerce, 619-852-4524.

## What to See in Lone Pine, California

▶ **The Alabama Hills** (off Whitney Portal Road). The hills are a series of rock formations located between this tourist town of two thousand and the nearby Sierra Nevada Mountains. The area has been used as a movie location since 1920. It's on public land operated by the Bureau of Land Management. In addition to *Death Valley Days*, shows that have gone on location here include *Bonanza, Have Gun Will Travel, The Lone Ranger, The Virginian, Wagon Train,* and *The Adventures of Wild Bill Hickok. The Twilight Zone* showed the moonscape location as a backdrop, and *The Rockford Files* filmed an episode in the town of Lone Pine itself.

▶ **The Lone Pine Film Festival.** Held annually in October, around the Columbus Day weekend, the festival includes tours of local TV and movie locations. Area businesses display historic photos from the shows and films, and plenty of souvenirs are offered for sale. For information on the festival, call 619-876-4314.

## Fun Facts

• In his autobiography, *An American Life,* Ronald Reagan says the job on *Death Valley Days* left him "plenty of time for speeches and Republican activities." During the show's run, Reagan headed Barry Goldwater's 1964 presidential campaign in California. After the election, he continued working on the show until the spring of 1965, when he was urged to run for governor.

- According to William Woodman, son of series creator Ruth Cornwall Woodman, his mother was not a big Ronald Reagan booster and was sorry *Death Valley Days* had helped Reagan gain national attention.
- Borax is extracted from an open mine pit by blasting, according to California-based U.S. Borax. The material is dissolved, crystallized, separated, and dried. Industrial borates are used, among other things, to make ceramics and glass. Twenty-mule teams really were used at one time to haul boron-containing ore.
- *Death Valley Days* ran on the airwaves for more than forty years, including its radio days, making it the western that was broadcast the longest.

## GOMER PYLE, U.S.M.C.
**(September 1964–September 1969, CBS)**

"Gollllleeee!"

Gomer (Jim Nabors) appeared on *The Andy Griffith Show* as a gas station attendant in Mayberry, but he decided to sign up for a five-year hitch in the U.S. Marine Corps and got his own military sitcom show. On *Gomer Pyle, U.S.M.C.*, Gomer was assigned to Camp Henderson in California, where he was a naive thorn in the side to gruff Sergeant Carter (Frank Sutton). But Gomer eventually won his superior over.

The Camp Henderson of the show existed mostly on a Hollywood set, but some location work was done at the real Marine Corps Recruit Depot on Henderson Avenue in San Diego, California, which trains recruits for the western half of the United States (its counterpart is on Paris Island, South Carolina).

### What to See in San Diego, California

▶ **Marine Corps Recruit Depot, Western Recruiting Region,** 1600 Henderson Avenue. Gomer Pyle (Jim Nabors) was seen walking through the archway on the parade grounds and elsewhere around this real base. In 1986, the base opened a museum in a former barracks building to help educate both the public and new recruits about the military branch. Included in the collection are posters, not from *Gomer Pyle, U.S.M.C.*, but from another military sitcom, *Major Dad*. The museum is open Tuesday to Saturday, 10 A.M. to 4 P.M. Call 619-524-6038.

### Fun Facts

- In real life, Jim Nabors, who was born in Sylacauga, Alabama, had a rich baritone singing voice that didn't at all suit his on-air bumbling Gomer demeanor. Nabors followed *Gomer Pyle, U.S.M.C.*, with *The Jim Nabors Hour* (September 1969 to May 1971 on CBS), an hour-long variety show, on which he got to sing.
- Frank Sutton, who played Sergeant Carter, and Ronnie Schell, who played Private Duke Slater on *Gomer Pyle, U.S.M.C.* also appeared on *The Jim Nabors Hour.*
- Jim Nabors also sang on sixteen albums including five gold and one platinum. He was a popular performer in Las Vegas and Lake Tahoe.
- Ted Bessell, who played Private Frankie Lombardi, went on to play Donald on *That Girl*.

## PETTICOAT JUNCTION
**(September 1963–September 1970, CBS)**

## GREEN ACRES
**(September 1965–September 1971, CBS)**

Hot on the heels of the success of the *The Beverly Hillbillies*, producer/writer Paul Henning created *Petticoat Junction*, a rural comedy starring Bea Benaderet (who earlier played Cousin

Pearl on *The Beverly Hillbillies*) as Kate Bradley, the widowed owner of the Shady Rest Hotel in the town of Hooterville.

Bradley had three gorgeous daughters, Billie Jo (Jeannine Riley, Gunilla Hutton, Meredith MacRae), Bobbie Jo (Pat Woodell, Lori Saunders), and Betty Jo (Linda Kaye), who helped her out along with the hotel's manager, Uncle Joe (Edgar Buchanan).

The critics hated the show; a *New York Post* review in 1963 said it had "numerous overtones . . . basically of the bathroom-plumbing, rural-sex variety . . . mostly, unnecessary." *Newsday* said it was "so bad that it may be popular . . ." Naturally, the show became a tremendous hit. And it was followed by *Green Acres*, also based in Hooterville.

*Green Acres* starred Eddie Albert as Oliver Wendell Douglas, a successful Harvard-educated New York lawyer who decided to join the back-to-earth movement, leaving the big city, and a Park Avenue abode, for a 160-acre farm in rural Hooterville. Douglas's glamorous wife, Lisa (Eva Gabor), was less than amused with Oliver's newfound love of farm life, but tried her best to get into the country ways. She burned pancakes while dressed in negligees and jewels.

Given that they lived in the same town, the characters of *Green Acres* and *Petticoat Junction* occasionally crossed paths on the shows, in what producer Henning called "cross-pollination."

Both shows were filmed mostly in Hollywood, but also did occasional location shoots in Jamestown, California, where a real steam engine, once used for hauling logs, was reborn as the *Cannonball* steam engine, which ran through Hooterville.

Henning said there was, to his knowledge, no real Hooterville, nor was there a real Shady Rest Hotel that inspired the series.

The aerial shot of the barn roof in the intro to *Green Acres* was filmed in Westlake Village, California, then a rural farm area, and now a bustling suburb of L.A. The barn is no longer there.

## What to See in Jamestown, California

▶ **Railtown,** 5th Avenue, at Reservoir Road. This twenty-six-acre state historic park has steam and diesel trains on display for visitors to view. The No. 3 train is the one that was used on *Green Acres* and *Petticoat Junction.* It was built in 1891, and like the other trains in the collection, the No. 3 was once the property of the Sierra Railway. Railtown has been used extensively as a location for TV shows, including *Lassie* and *Little House on the Prairie,* and movies. The park grounds are open daily, and guided tours are offered from 9:30 A.M. to 4:30 P.M. A small fee is charged. Steam train rides are also offered on weekends. Call 209-984-3954.

## Fun Facts: *Petticoat Junction*

- Bea Benaderet was also the original voice of Betty Rubble in *The Flintstones.*

The Number 3 train at Railtown. *(Photo by Larry Ingold, courtesy of Sierra Railway Company of California, Jamestown, California.)*

• Actress Linda Kaye, who played Betty Jo, is creator/executive producer Paul Henning's daughter. She was the only one of the original "Jo" daughters who stayed with the show.
• When actress Benaderet died in 1968, June Lockhart *(Lassie)* joined the cast as Dr. Janet Craig, the town's new doctor.

**FUN FACTS: *GREEN ACRES***

• Lisa (Eva Gabor) and Oliver (Eddie Albert) met when he was an American pilot in World War II and she was in the Hungarian underground.
• The theme song featured vocals by Gabor and Albert. He: "Green Acres is the place to be . . ." She: "New York is where I'd rather stay . . ."
• Gabor, who earlier appeared on talk shows and on Broadway in *The Happy Time,* was born in Budapest and is the younger sister of Zsa Zsa, as well as Magda Gabor.
• Actor Albert is from Minnesota; he starred on Broadway in *Brother Rat.*
• Gabor often wore her own jewelry on the show, but it was all fake. In real life, as on the show, she was known as a lousy cook.
• The characters in the Douglas's adopted hometown included Arnold, the adopted son of Fred Ziffel (Hank Patterson), who happened to be a television-watching pig.
• In a 1967 episode, the Hooterville residents did a charity show based on *The Beverly Hillbillies.* Oliver (Eddie Albert) played Jethro, Lisa (Eva Gabor) was Granny, and Hank Kimball (Alvy Moore) was Jed.

## FYI: ALSO IN CALIFORNIA

The short-lived musical series *Seven Brides for Seven Brothers* (September 1982 to July 1983, CBS) starred actors Richard Dean Anderson *(MacGyver)* and Peter Horton *(thirtysomething),* as well as a young River Phoenix (who later died from a drug overdose).

The show filmed on location in Calaveras County, including Angels Camp, home of an annual frog jump, which was made famous by writer Mark Twain in "The Celebrated Jumping Frog of Calaveras County."

## What to See in Calaveras County, California

- **Murphys Historic Hotel & Lodge,** 457 Main Street, Murphys. This twenty-nine-room registered historic landmark property was featured in *Seven Brides for Seven Brothers,* as well as in an episode of *Highway to Heaven.* Stars from both shows stayed at the property, which also has a restaurant and an old-time saloon. Call 209-728-3444.

- **Frogtown, Angels Camp.** Home of the International Jumping Frog Jubilee, held annually the third weekend in May, this county fairground also hosts a variety of agricultural, business, social, and educational events during the year. The pilot for the CBS western series *Paradise* was also filmed here. Call 209-736-2561.

For more information on Calaveras County and California's Gold Country, call the Calaveras Lodging & Visitors Association, 800-225-3764.

# SAN FRANCISCO AND ENVIRONS

The San Francisco Police Department has been the subject of several shows including *The Lineup* (also called *San Francisco Beat*) in the 1950s, which was based on real SFPD cases; the popular *Ironside*, about a paralyzed former SFPD chief of detectives; *The Streets of San Francisco*, about a team of SFPD cops; and *Midnight Caller*, about a former SFPD cop turned radio talk show host. All the shows shot scenes in the Bay City, with *The Streets of San Francisco* and *Midnight Caller* shooting entirely on location here.

A real row of historic San Francisco homes, with scenic views in the background, was seen in the intro to the family sitcom, *Full House*, which had a plot based in the Bay City, but which was really shot in L.A.

Other sitcoms that have also called San Francisco home, at least in their plots, include *My Sister Sam*, *Too Close for Comfort*, and *Phyllis*.

*Hangin' with Mr. Cooper* is supposed to take place in nearby Oakland, California.

The real-life landmark Fairmont Hotel on Nob Hill was the setting seen as the posh St. Gregory on *Hotel*. And *Trapper John, M.D.* (a spin-off of *M*A*S*H*) showed city sites including Ghirardelli Square and Fisherman's Wharf.

Napa Valley, California's scenic wine country just to the north of San Francisco, was the setting for the soap opera goings-on at *Falcon Crest*, with real-life vineyards featured.

Up the coast (about 156 miles) is the very quaint coastal village of Mendocino, the real-life seaside setting seen on *Murder, She Wrote* (although it's supposed to be Cabot Cove, Maine, on the show).

South of San Francisco, in Woodside, California, is the mansion (in real life, a museum) seen as the wealthy Carrington family's estate on *Dynasty,* although on that nighttime soap the setting was supposed to be Denver.

## FALCON CREST
**(December 1981–May 1990, CBS)**

Known at first as the show that followed *Dallas, Falcon Crest,* with its plush northern California winery vistas, drew its own loyal fans, who enjoyed following the theatrics of villain Angela Channing (Jane Wyman) and the other residents of fictional Tuscany Valley, California. Power, sex, and money were all integral to the plot of this nighttime soap opera, but so was wine making.

Angela runs the Falcon Crest Winery (which she eventually loses after much power-playing shenanigans, and later wins back).

Angela's nephew, Chase (Robert Foxworth), the show's good guy, has his own fifty acres, inherited from his deceased father, much to Angela's displeasure. And creepy publisher Richard Channing (David Selby) also eventually gets into the wine scene (it turns out he's really Angela's son, whom she thought was stillborn).

At the end of the series, the Falcon Crest Winery was to be passed to a new generation. And when the show went off the air, it ended with the toast, "The land endures."

While much of the show was filmed in Hollywood, *Falcon Crest* also did a lot of location shooting in real California wine country, using the city of Napa, California, as well as Yountville, Rutherford, and St. Helena, for background footage.

The main winery seen on the show was the real Spring Mountain Vineyards, in St. Helena. Unfortunately, under its current ownership, the winery is not open to visitors. Nor is the mansion visible from the road.

Richard Channing (Selby) lived in what is in real life the house at the Altamura Winery in Napa, but that too is not open to the public, nor is it visible from the road.

But Inglenook Vineyards, which was also featured on the show, welcomes visitors and has staff ready and willing to talk about vineyard life and explain the wine-making process.

## What to See in Rutherford, California

▶ **Inglenook Vineyards,** 1991 St. Helena Highway. The stained-glass-windowed barrel house of this winery, one of the oldest in Napa Valley, appeared on the show. And a *Falcon Crest* wedding scene was filmed in the winery's courtyard. Fifteen different varieties of wine, under the Inglenook and Gustave Niebaum labels (that's the name of the winery's founder), are produced here. The winery offers free tours, which include the stone cellars built into the hillside, the vineyards, and the wine-aging facility (which holds 2,200 barrels used to age red wine), as well as tastings at the hospitality center. There is also a retail store. The site is now operated as **Niebaum-Coppola Vineyards, Estate and Winery.** Tours are offered by appointment only. Call 707-963-9099.

For visitor information in Napa Valley, call the Napa Valley Conference & Visitors Bureau, 707-226-7459.

## Fun Facts

- Actress Jane Wyman, who played Angela, was once Mrs. Ronald Reagan. They married in 1940, and divorced in 1948.
- *Falcon Crest* was originally to be called *Vintage Years,* which was also the name of the show's pilot.
- The series was created by Earl Hamner, whose youth was re-created in his earlier show, *The Waltons.*
- Ken Olin *(thirtysomething)* appeared on *Falcon Crest* in the role of Father Christopher.
- Rock singer Patricia "Apollonia" Kotero, an often scantily clad protégée of rock star Prince, played Apollonia on the show.
- During the show's run, the real Spring Mountain Vineyards offered a "Falcon Crest" label wine, in honor of the show.
- *Falcon Crest* featured movie actors and actresses as guest-stars, some in rare TV appearances, including Lana Turner, Gina Lollobrigida, Kim Novak, Cliff Robertson, and Cesar Romero.
- The show created a new sex symbol in actor Lorenzo Lamas, who played Lance Cumson, Angela's playboy grandson. The actor was tapped to host *Dancin' to the Hits,* a short-lived syndicated rock-and-roll dance show.
- In 1989, Lorenzo Lamas married actress Kathleen Kin-

mont. The two starred together in the syndicated action-adventure series *Renegade*, introduced in 1992. They continued to work together despite their separation in 1993.

## FULL HOUSE
**(September 1987–May 1995, ABC)**

Comedian Bob Saget starred in this family sitcom as widower Danny Tanner, a TV anchor who raised his three adorable and fast-growing daughters with the help of his Elvis fan brother-in-law, Jesse Cochran/Katsopolis (John Stamos), and boyish friend Joey Gladstone (David Coulier).

All-too-cute fraternal twins Mary Kate and Ashley Olsen shared the part of the youngest daughter, Michelle. Later, Jesse got married to Rebecca (Lori Loughlin), and they had twins, Alexander and Nicholas (Blake and Dylan Wilhoit), adding another cute set to the cast.

The other daughters were D.J. (Candace Cameron), who blossomed as a teenager before viewers' eyes, and middle child Stephanie (Jodie Sweetin), who complained about her sisters with comments like "How rude!"

The show was set in San Francisco, and the exterior shown as the Tanner's home is in the city's historic Alamo Square. *Full House* was actually shot in a studio in Burbank, however. The show was rapped a bit for its sappiness, but it occasionally dealt with real issues such as a poignant episode in which Michelle learned to deal with the death of a loved one.

*Full House* was extremely popular with children and teenagers and was a hit show not only in prime time but also in syndication.

### What to See in San Francisco, California

- **Alamo Square row houses** (on Steiner, between Grove and Hayes). The Victorian row houses in this residential neighborhood, featured in exterior shots on the show, are also referred to as the "Painted Ladies." The skyline of the city can be seen rising behind the homes, which have appeared in several films as well. The nearby Alamo Square Park has tennis courts and jogging trails as well as green areas for relaxing, but it's also in

The Victorian houses at Alamo Square. *(San Francisco Convention & Visitors Bureau photo, by Carol Simowitz.)*

an inner-city area, and walking at night in the neighborhood is not advised.

## Fun Facts

- Scott Weinger, who played D.J.'s boyfriend, Steve, attended Harvard in real life, and played the title character's voice in the Disney movie *Aladdin*. He also reported on youth issues for *Good Morning, America*.

• While doing *Full House,* hardworking Bob Saget also hosted *America's Funniest Home Videos* on ABC. Saget started his career as a stand-up comic.

• Teenage heartthrob John Stamos, who played Jesse, earlier played Blackie Parrish on *General Hospital,* and won an Emmy for the role. In real life, Stamos, like his character, is of Greek descent, and his last name is a shortened version of Stamotopoulos.

• Mary Kate and Ashley Olsen, who shared the role of Michelle, recorded an album of children's songs in 1992, which they promoted to the hilt.

• Kimmy Gibler (Andrea Barber) was D.J.'s best friend, and the character had extraordinarily smelly feet.

• The show was produced by the same team that did *Family Matters* and *Step by Step,* Tom Miller and Bob Boyett.

## HOTEL
**(September 1983–August 1988, ABC)**

Romance, drama, adventure, excitement, and intrigue happened within the walls of the stately St. Gregory Hotel in San Francisco, with a different cast of guest stars each week interacting with the hotel's staff. Given the upscale surroundings, can you tell *Hotel* was produced by the same people who did *The Love Boat?* There's no Captain Stubing here, but there was a bearded James Brolin, who starred in *Hotel* as Peter McDermott, the general manager of the St. Gregory.

Movie star Bette Davis starred in the show's premiere, as Laura Trent, the hotel's aristocratic owner. But when Davis suddenly became ill she was replaced by another veteran actress, Anne Baxter, who appeared as Trent's sister, Victoria Cabot, overseer of the operation in Laura's absence.

Actress Baxter died in 1985, and her death was reflected in the story line, with McDermott inheriting part of the property. The other half went to a scheming bunch of Cabot's relatives, including Charles (Efrem Zimbalist, Jr.), Jessica (Dina Merrill), and Jake (Ralph Bellamy). McDermott eventually managed to become the full owner, and his lovely assistant Christine (Connie Sellecca) became the general manager of the hotel.

The Fairmont Hotel. *(Photo courtesy of the Fairmont Hotel, San Francisco.)*

The real-life posh property seen on the show is the Fairmont Hotel on Nob Hill. The show's pilot was actually shot on location at the Fairmont, using the property's exterior as well as the lobby and the Benjamin H. Swig Presidential Suite, according to the hotel's former managing partner, Rick Swig.

Exteriors of the real property, including the front door, were also shown on every subsequent episode, and the lobby was duplicated on the show's Hollywood set, Swig said. The re-creation was so real that Swig said his mom, forgetting she was not in the real hotel, tried to give someone directions to the hotel's bathroom while visiting the set.

## WHAT TO SEE IN SAN FRANCISCO, CALIFORNIA

▶ **Fairmont Hotel,** atop Nob Hill. This landmark property featured on *Hotel* has housed visiting dignitaries and members of high society since 1906. The interior decor is opulent European, including a grand staircase. *Murder, She Wrote* is one of many other TV and film productions that have visited the property. The hotel has 600 luxury guest rooms, with rates beginning at about $175 a night. The eight-room penthouse suite featured on the show rents for $6,000 a night. The hotel also has five restaurants offering an array of cuisines. For reservations, call 800-527-4727.

### Fun Facts

• James Brolin earlier appeared on *Marcus Welby, M.D.*, and won an Emmy for his supporting role as Dr. Steven Kiley on the show.

• Guest stars on *Hotel* included Morgan Fairchild, Mel Torme, Shirley Jones, Pernell Roberts, Liberace, and Lynn Redgrave.

• The show was based on Arthur Hailey's best-seller *Hotel.* And the show was also known as *Arthur Hailey's Hotel.*

• Heidi Bohay, who played desk clerk Megan Kendall, and Michael Spound, who played bellhop Dave Kendall, Megan's husband, were married in real life too.

• Shari Belafonte-Harper, who played hotel employee Julie Gillette, is also a model and the daughter of singer Harry Belafonte.

• Singer-turned-actress Michelle Phillips appeared as Elizabeth Bradshaw Cabot on the show. She later joined the cast of *Knots Landing.*

• Connie Sellecca, who played Christine, earlier appeared in *The Greatest American Hero.*

## IRONSIDE

**(September 1967–January 1975, NBC)**

After nine years on *Perry Mason,* actor Raymond Burr jumped into another successful and long-running show, starring as Robert Ironside, the crusty former chief of detectives of the San Francisco Police Department (SFPD). Ironside was paralyzed by an assassin's bullet and assists the SFPD on a special consultant basis. The wheelchair-bound Ironside investigates around

San Francisco's Ferry Building. *(San Francisco Convention & Visitors Bureau photo, by James Martin.)*

town in a specially equipped van, with the help of an able team of assistants. The show was hailed as an inspiration by advocacy groups for the handicapped.

According to Lieutenant Dennis Chardt of the SFPD, *Ironside* did some location shooting in San Francisco, where the plot is based, including, he believes, at the Ferry Building on the waterfront. (Officials at the building, however, said they have no record of filming there). The historic one-time agricultural building at the foot of Mission Street at the Embarcadero may also have been used, Chardt said. However, the show was mostly filmed in Hollywood.

### What to See in San Francisco, California

▶ **The Ferry Building** (World Trade Center), at the foot of Market Street (at the waterfront). This historic building is home to the Port of San Francisco offices as well as other offices, and is the departure point for the Golden Gate Ferry and the Red and White Ferry. The green sandstone building, with its large clock tower, was built in 1896.

### Fun Facts

• Raymond Burr, who died in 1993 after losing a battle with cancer, was a known workaholic. He appeared in ninety feature films including Alfred Hitchcock's *Rear Window* before becoming Perry Mason on TV.

• Actor Burr's other roles included a reporter in the original *Godzilla* movie.

• NBC brought the character of Ironside back in 1993, with the TV movie *The Return of Ironside*. Joining Burr in the movie reunion were Ironside's original team including Detective Ed Brown (Don Galloway), policewoman Eve Whitfield (Barbara Anderson), policewoman Fran Belding (Elizabeth Baur), and personal assistant Mark Sanger (Don Mitchell).

## MIDNIGHT CALLER
**(October 1988–August 1991, NBC)**

Jack Killian (Gary Cole) was a cop who left the force and became a radio talk show host, after accidently killing his partner, on this drama series. The socially conscious Jack continues to solve crimes and other problems, responding to calls from listeners of his San Francisco radio show, *Midnight Caller.* The show was shot entirely on location in the Bay Area, including scenes in North Beach, the city's lively Italian neighborhood.

The real house seen as Jack's on the show, according to local sources, is near Alamo Square, at the corner of Steiner and Fulton. Later Jack moved into a loft, with exteriors shown of a commercial building at Lombard and the Embarcadero.

Other locations seen on the show included Alcatraz prison, which was featured in a prison riot episode.

### What to See in San Francisco, California

▶ **North Beach.** This neighborhood is one of the oldest in the city and is known as the Little Italy of the West. The show shot scenes in Washington Square (where public t'ai chi classes take place early in the morning) and featured the steps of the Romanesque Saints Peter and Paul Church (666 Filbert Street). The narrow alleyways on the sides of Telegraph Hill also appeared on the show, as did the restaurants on upper Grant Avenue. This colorful area includes Lombard Street, known as the crookedest street in the world. Restaurants and Italian cafes abound.

### Fun Facts

- G. Gordon Liddy, the Watergate conspirator, guest-starred on an episode of the show as a mysterious power broker who tried to meddle in an important San Francisco city election.
- Gary Cole earlier appeared in the TV miniseries *Fatal Vision.*
- Actor Peter Boyle guest-starred as Jack's (Cole) con-artist dad, J.J.
- The show was honored by the city of San Francisco in 1991, with a proclamation of the mayor declaring a *Midnight Caller* Day in the city.
- Wendy Kilbourne, who played KCJM station owner Devon King, left the show in 1990 to have a baby. Her pregnancy was written into the story line, with King deciding to sell the station due to her own pending motherhood.
- Lisa Eilbacher, who played Nicky, the new station manager, was married in real life to Bradford May, executive producer of another San Francisco-based show, *Over My Dead Body,* on CBS.
- Actress Eilbacher earlier appeared in the movies *Beverly Hills Cop* and *An Officer and a Gentleman.*

• Betty Thomas and Ed Marinaro, of *Hill Street Blues*, were reunited on *Midnight Caller* as guest stars.

• Actress Kay Lenz played Jack's ex-girlfriend, who was dying from AIDS. Lenz, who won an Emmy for her *Midnight Caller* role in 1989, was once married to David Cassidy *(The Partridge Family).*

## MURDER, SHE WROTE
**(September 1984– , CBS)**

Stage and film actress Angela Lansbury became an even bigger star playing Jessica Fletcher, a sweet middle-aged eccentric mystery novelist turned amateur sleuth in this popular detective show. The character was much loved by audiences.

Fletcher (Lansbury), as fans know, hails from the quaint town of Cabot Cove, Maine. And even when she moves to New York on the show in 1991, to teach criminology part-time at Manhattan University (some scenes were actually filmed at UCLA), the character continues to weekend in her hometown.

But Cabot Cove is fictional. And while the setting may look like New England, the town seen on the show is really Mendocino, California, the northern California seaside village. (That's the Pacific Ocean, not the Atlantic, you can see in the background.)

The move of Jessica from Cabot Cove to New York coincided with Lansbury herself taking over as executive producer of *Murder, She Wrote*. Of the move, Lansbury told *The New York Times*, "The New York locale just allows us to introduce a far wider ethnic diversity than Maine."

Another reason for the move, said a spokeswoman for Universal Television, was that the show's scripts often involve murders, and everybody was dying in tiny Cabot Cove! The writers needed to have a larger group of people to interact for plot lines.

Jessica does a lot of traveling on the show, mostly to visit friends and do book tours, and there always seems to be a mystery for her to resolve on the road, as well. While a lot of the travel settings have been re-created in a Hollywood studio, where most interiors for the show are shot, episodes of the show have been filmed on location in Ireland, where Lansbury once lived, and in Hawaii. And a crew was sent to Hong Kong for footage for one story line.

*Murder, She Wrote* is not the only TV project to have filmed in Mendocino, California. *The F.B.I.* shot an episode on location here in 1971 called "Bitter Harbor in Mendocino," and a lot of movies have been filmed here too, such as *Dying Young, Same Time Next Year,* and *Racing With the Moon.*

## WHAT TO SEE IN MENDOCINO, CALIFORNIA

▶ **Blair House,** 45110 Little Lake Street. The exterior of this Victorian bed and breakfast, built in 1888, appears as Jessica's Cabot Cove home on the show. The property offers four rooms, including one two-room suite, in the main house and a cottage out back. Guests are served breakfast at the nearby Mendocino Bakery. Room rates are $75 to $130 per room, per night, including breakfast. The bed and breakfast sells postcards that indicate its use on the show. Call 707-937-1800.

▶ **Hill House Inn,** 10701 Palette Drive. Both exteriors and interiors of this real inn appear on the show. The establishment's sign out front had both Hill House Inn, Mendocino, and Hill House Inn, Cabot Cove, printed on it, so the sign could be flipped on shooting days. Located about two blocks from the town's downtown area, the inn has forty-four rooms, as well as a restau-

The Hill House in Mendocino, California. *(Photo courtesy of The Hill House.)*

rant featuring country French cuisine and serving breakfast, lunch, and dinner. Rates are $110 to $175 per room, per night. In addition to filming here, the cast and crew stayed at the inn when they were in town (Angela Lansbury has slept here!), and pictures of the cast can be seen in the lobby. Other celebrities who have stayed here include actress Bette Davis, and there is a suite named in her honor. Call 707-937-0554.

## Fun Facts

• As a stage actress, Angela Lansbury won four Tony awards for *Mame, Dear World, Sweeney Todd,* and *Gypsy.*

• As a movie actress, Lansbury was nominated for Oscar awards for *Gaslight* (in which she appeared at age eighteen), *The Picture of Dorian Gray,* and *The Manchurian Candidate.*

• While nominated more than a dozen times for an Emmy, Lansbury had not won one as of 1994. But she did get to host the Emmy telecast in 1993.

• Actress Lansbury became famous to younger fans as the voice of the teapot in Disney's *Beauty and the Beast.*

• In a 1991 episode, a postal carrier drops off a letter and asks for postage due. The actor is none other than U.S. Postmaster General Anthony Frank, making a cameo appearance.

• Among fans of *Murder, She Wrote* were President George Bush and his wife, Barbara, who invited the English-born Lansbury to White House dinners. The Bushes, like Jessica, also have a little place in Maine (in their case, an oceanfront mansion in Kennebunkport).

• Among the places Jessica has visited on the show are New Orleans, Boston, New Mexico, Hawaii, Dallas (where she recuperated from a broken leg), London, Idaho, Las Vegas, San Francisco, Washington, D.C., Paris, Seattle, Athens, Palm Springs, Monte Carlo, Hong Kong, and Hollywood (including a stop at the *Psycho* House at Hollywood's Universal Studios, to investigate a murder there).

• Producing *Murder, She Wrote* is a family affair. Lansbury's husband, Peter Shaw, is her manager. Stepson David, heads her production company, Corymore. Anthony, her son by Peter, has directed many episodes of the show. And Lansbury's younger brother, Bruce, is a supervising producer.

• Lansbury is a second-generation actress. Her mother, Moyna MacGill, performed in London's West End in the 1920s and 1930s.

• A frequent guest on *Murder, She Wrote* was Tom Bosley *(Happy Days)*, who played Cabot Cove Sheriff Amos Tupper. Bosley left the show to do his own mystery-solving show, the *Father Dowling Mysteries.*

## THE STREETS OF SAN FRANCISCO
**(September 1972–June 1977, ABC)**

Shot on location, *The Streets of San Francisco* showed much of the city as the haunts of Lieutenant Mike Stone (Karl Malden) and Inspector Steve Keller (Michael Douglas) as they fought crime for the San Francisco Police Department. On the show Stone was a tough veteran cop, brought up through the ranks, and Keller was his young, college-educated partner. The two worked for the Bureau of Inspectors Division of the SFPD.

Real police buildings appeared on the show, but the police headquarters seen was really the San Francisco Hall of Justice, a courthouse building. Stone and Keller were often seen walking and talking in a breezeway at the building, and a police car was seen leaving a back parking lot at the Hall of Justice in the show's intro.

*The Streets of San Francisco* also had permanent sets in a warehouse below Telegraph Hill.

### WHAT TO SEE IN SAN FRANCISCO, CALIFORNIA

▶ **San Francisco Hall of Justice,** 850 Bryant Street. This courthouse building was used as the police station on the show. The parking lot where the police car pulled out in the show's intro is no longer there (it's given way to an addition to the building). Some scenes were also filmed inside the building, according to police insiders.

▶ **San Francisco City Hall,** 400 Van Ness Avenue. The city hall appeared as itself on the show. In real life it's home to the mayor's office and other city offices. The domed building was

San Francisco's City Hall. *(San Francisco Convention & Visitors Bureau photo, by Judy Houston.)*

dedicated in 1915, and is of French Renaissance design. The interior includes California marble.

## Fun Facts

• Michael Douglas, son of veteran actor Kirk Douglas, later romped with another Stone, actress Sharon, in the movie *Basic Instinct.* His successful film career also included *Wall Street* (for which he won an Oscar) and *Romancing the Stone.*

• *The Streets of San Francisco* was based on the novel *Poor, Poor Ophelia,* by Carolyn Weston.

• Richard Hatch joined the show in 1976 as Inspector Dan Robbins (after Douglas departed). Hatch later appeared on *Battlestar Galactica.*

• Robert Wagner *(Hart to Hart)* was a guest star in the premiere episode of *The Streets of San Francisco,* playing a young attorney whose business card was found on a murdered woman.

• Actor Karl Malden later hosted some early episodes of *Unsolved Mysteries*, before that show was signed on as a regular NBC series.

## FYI: ALSO IN SAN FRANCISCO AND ENVIRONS

The producers of *Dynasty* (January 1981 to May 1989, ABC) snubbed Denver by declaring they could not find a suitable posh mansion in the Rocky Mountain city to show as the house of the wealthy Carringtons. Instead, they chose a California landmark, the Filoli Mansion in Woodside, California, which is operated as a museum by the National Trust for Historic Preservation.

The show filmed scenes inside and outside the mansion, as well as in the formal gardens. And four rooms of the house, the front hall, library, living room, and reception room, were later re-created on the show's Hollywood set.

The Filoli Mansion. *(Photo courtesy of the Filoli Center, Woodside, California.)*

A rear view of the Filoli Mansion. *(Photo courtesy of the Filoli Center, Woodside, California.)*

## What to See in Woodside, California

- **Filoli Mansion,** off Highway 280 (about thirty miles south of San Francisco). The mansion was built in 1916 for William Bower Bourn, Jr., owner of the Empire Mine, a gold mine in Grass Valley, California. The name of the mansion is derived from the motto "To *fi*ght for a just cause, to *lo*ve your fellow man and to *li*ve a good life." The mansion has also appeared in several movies. The museum is open from mid-February to the first week in November. Tours are offered of the mansion and its seventeen acres of formal gardens on a reservations-only basis, Tuesday, Wednesday, and Thursday at 10 A.M. and 1 P.M., and on Saturday, every half hour from 9:30 A.M. to 1:30 P.M. On Friday, visitors can tour the mansion and grounds on a self-guided basis, without reservations. Call 415-364-2880.

(See also the West.)

# THE PACIFIC NORTHWEST

Perry Como sang, "The bluest skies you've ever seen are in Seattle . . ." So it's really no wonder the state of Washington would attract TV productions. The song, in fact, comes from the 1960s series *Here Come the Brides*, which has a plot based in 1870s Seattle.

But it wasn't blue skies that movie director David Lynch was seeking for his unusual series *Twin Peaks*. It was more the dark and eerie mood that he saw in the abundance of Douglas firs in and around Snoqualmie, Washington, the real-life setting seen as the small town of Twin Peaks on the show. Actually, buildings and settings in several towns located in a half-moon area about a half hour's drive east of Seattle appeared on *Twin Peaks*, including a real hotel that doubled as the show's Great Northern Lodge, and a real cafe that appeared as the Double R Diner. The twin peaks seen on the show are part of the real Mount Si. *Twin Peaks* mainly shot exteriors in the area, with the show's actors doing most of their work on a set in L.A.

*Northern Exposure* was the first show to shoot entirely on location in Washington. The real-life historic coal-mining town of Roslyn, Washington, about eighty-five miles east of Seattle (like Snoqualmie, located off I-90) in the Cascade Mountains, was the setting seen on *Northern Exposure*. But on the show, Roslyn doubled for fictional Cicely, a town located in Alaska, not Washington. Practically the whole town of Roslyn appeared on the show, and will look familiar to *Northern Exposure* fans.

## NORTHERN EXPOSURE
**(July–August 1990, April 1991–July 1995, CBS)**

Young Dr. Joel Fleischman (Rob Morrow), who is from Flushing in Queens, New York, paid off his medical school bills by serving as the physician in tiny and faraway Cicely, Alaska. There he befriended an odd bunch of off-beat characters that included Maggie (Janine Turner), a bush pilot whose boyfriends always seemed to mysteriously die.

Joel initially missed the big city, but stayed in town long past the time he paid off his debt, using the remote location, gorgeous setting, and cool air for some philosophical thinking about his religion (Jewish) and life in general.

The doctor also moved past studying in an academic manner the town's other residents. And as the years went by, he seemed to fit in more and more with the hearty Alaskan clan.

The scenic setting seen on the show as Cicely was not really Alaska, but it's not Hollywood either. *Northern Exposure* was the first TV show to shoot entirely on location in the state of Washington, and the real-life town seen on the show was Roslyn, Washington, population 875.

Cicely, the show's fictional Alaska town, was founded by a lesbian couple, Cicely and Roslyn. (The pairing helped the producers explain the Roslyn signs that sometimes appeared in shots of the real town of Roslyn). In real life, Roslyn was founded at the turn of the century, as a coal-mining town.

The whole main street of Roslyn, Pennsylvania Avenue, appeared on the show. And most locations seen were real buildings or homes (an exception being Joel's cabin, which was a pretend house located on private property). Interiors for the show were done on a soundstage in Redmond, Washington, right outside of Seattle.

Charles Carroll, unit production manager for the show, said Roslyn was chosen partly because filming in a real Alaska town was ruled out as too costly. Roslyn was the nearest town to Seattle (and L.A.) with "a rustic Alaskan kind of flair to it. It was almost like a ghost town with old buildings in the mountains," Carroll said.

To further create an Alaskan ambiance in Roslyn, the crew added Native American totem poles on Pennsylvania Avenue, and affixed deer antlers on some buildings, according to J. Daniel Dusek, location manager for *Northern Exposure.* The crew also added new paint to some buildings, and some signs in the town that didn't quite fit in were removed or covered up on shooting days.

The snow seen on the show was often real, adding a frigid realism to the setting. (The actors didn't always have to pretend to be cold!)

The town of Roslyn, very aware of the tourist status it gained as home to *Northern Exposure,* renumbered Pennsylvania Avenue to make it easier for visitors to find locations used on the show. Kurt Peppard, a Gray Line bus driver, also created a tour book of Roslyn, for sale at local gift shops, with part of the proceeds going to town projects.

The show meant a boon for the tiny town, not just in tourism, but also in real estate prices (they've risen). And most residents of Roslyn seemed to like the town's TV status (many appeared as extras on the show). Some complained publicly, however, about their little town being invaded by Hollywood.

## What to See in Roslyn, Washington

▶ **The Roslyn Cafe,** 201 W. Pennsylvania Avenue. The big mural seen on the show with the camel painted on it is really outside this cafe. On the show, the spot is named for Roslyn, who founded the fictional town of Cicely with her lover, Cicely. Open Tuesday to Sunday, 9 A.M. to past dinner. Call 509-649-2763.

▶ **The Brick Tavern,** 100 W. Pennsylvania Avenue (at Highway 903). In real life and on the show, this bar had the same name (that's the real neon sign viewers saw on the show). But on *Northern Exposure,* where it's operated by Holling (John Cullum), the place served meals. In real life, The Brick, which dates to 1889 and is the oldest continuously operating tavern in the state of Washington, serves only beer and wine and light snacks. Souvenirs of the show are also for sale. Open daily, noon to 2 A.M. Call 509-649-2643.

▶ **Roslyn Realty,** 112 W. Pennsylvania Avenue. This office appeared as Joel Fleischman's (Rob Morrow) office on the show. Across the street is Totem Pole Park, so dubbed by locals because the previously empty lot was where the show's crew planted totem poles to give the town an Alaskan ambiance.

Downtown Roslyn, Washington, known as Cicely, Alaska, on *Northern Exposure*. *(S. Breyfogle Photo.)*

▶ **Central Sundries,** 101 W. Pennsylvania Avenue. Ruth-Anne's (Peg Phillips) place on the show, this shop in real life offers *Northern Exposure* memorabilia, T-shirts, and other souvenirs. Also for sale are soft drinks and candy, office supplies, and liquor. Call 509-649-2210.

▶ **Memory Makers,** 101 E. Pennsylvania Avenue. This big souvenir shop is located in the old Northwest Improvement Company Store, a historic building that dates back to the time when Roslyn was a mining town (it's where the miners bought supplies, about a hundred years ago). On the show, part of the building appeared as KBHR Radio. Call 509-649-2557.

▶ **It's Country Plus** and **Illuminated Myst Bookstore and Espresso,** 101 N. 2nd Avenue. This building appeared as the Cicely Laundromat on the show. In real life, it houses a gift shop and a bookstore. Call 509-649-3663.

▶ **Cicely Gift Shop,** 103 N. 2nd Avenue. While it had not appeared on the show, this shop was obviously angling for a starring role. Souvenirs are for sale here. Call 509-649-3080.

▶ **The private houses on E Street.** If you ask local residents for directions, you may be able to catch a glimpse of the private homes that appeared on the show as Maurice's (Barry Corbin) and Maggie's (Janine Turner) cabins.

▶ **Lake Cle Elum,** Highway 903 (about six miles northwest of Roslyn), and **Salmon la Sac** (about thirteen miles up the same road). The gorgeous lake and river scenery seen on the show was often shot in this area. Highway 903 runs through the Wenatchee National Forest and offers a beautiful scenic drive.

## FUN FACTS

• Joshua Brand and John Falsey created the series. They also created *St. Elsewhere.*

• Janine Turner, who played Maggie, was briefly engaged to actor Alec Baldwin.

• Actress Turner and movie star Demi Moore were the Templeton sisters on *General Hospital* from 1982 to 1983. Turner played Laura Templeton and Moore played her sister, Jackie.

• Actor John Corbett, who played DJ Chris Stevens, became known in real life as a sensitive sex symbol as a result of his role on the show.

• *The "Northern Exposure" Cookbook: A Community Cookbook from the Heart of the Alaskan Riviera,* was published in 1993.

• On the show, Joel (Rob Morrow) went to Columbia University Medical School. He originally thought he would be working his tuition debt off in Anchorage, not the hinterlands.

• Actor Morrow wore L.L. Bean-type rugged clothing on the show, but in real life, the actor is often seen in Armani suits and the like, according to *People* magazine.

• Morrow decided to leave the show to pursue a movie career in 1994, after receiving good reviews for his performance in Robert Redford's *Quiz Show.* Morrow, who once studied photography, practiced his skills by taking candid shots of fellow cast members, and a collection of the photos was published in 1994 as *Northern Exposures,* by Warner Books.

• The population of fictional Cicely is 814, and it is on the Alaska Riviera.

• Holling (John Cullum) started dating Shelly (Cynthia

Geary) when she was eighteen and he was at least in his fifties. They married and had a baby daughter, Miranda.

• Joel (Morrow) never exactly hired Marilyn (Elaine Miles), his unemotional medical assistant. She just assumed the role.

• Maggie (Turner) and Joel (Morrow) did sleep together once. But both chose to forget it. It happened in February 1993.

• The pop music soundtrack of *Northern Exposure* was released as an album.

• The show featured on one episode the first fictional gay wedding (male) on network TV.

• The moose seen in the show's intro was reportedly "Morty the Moose," who was owned by Washington State University. He passed away after his TV debut, but not before siring at least four calves.

• The show was so popular that Holland America Westours and its subsidiary Gray Line of Alaska added a stop in Roslyn to some of their tour itineraries. On one episode of the show, a Gray Line of Alaska tour bus made a cameo appearance.

• According to *Advertising Age,* the show's unusual on-air promotion in November 1992 offering a *Northern Exposure* sweatshirt for sale sold thousands of sweatshirts, and also provided CBS and MCA/Universal Merchandising with a data base of upscale consumers who usually don't purchase such licensed products.

## TWIN PEAKS
**(April 1990–June 1991, ABC)**

Who killed Laura Palmer? That was a major mystery, among many, in this serial show, which cocreator David Lynch described as "a murder mystery–soap opera with fantastic characters." The answer is, Laura's (Sheryl Lee) father, Leland (Ray Wise), who also killed her cousin, Madeleine (Sheryl Lee), as his hair turned from black to gray. But Leland was possessed by evil spirit Bob (Frank Silva), so did it really count?

FBI Agent Dale Cooper (Kyle MacLachlan) came to the small town of Twin Peaks to search for Laura's killer and never left, finding the setting conducive to his own search for understanding of people's capacity for good and evil (including his own).

Cooper recorded the sexual antics, violence, and other oddities and evils he encountered among town residents on his handheld cassette player, presumably for his secretary, Diane, to transcribe someday. Among his advisers in his quest to understand good and evil were dancing dwarfs, a boy in a mask, and a lethargic giant.

The show's dark themes and eerie, slow pacing, highlighted by an unusual musical score, were much talked about among critics and viewers. And like daring filmmaker Lynch's earlier movie *Eraserhead,* the show became a cult classic, with a devoted following of fans. *Twin Peaks* never had the numbers of fans the network would have liked, however, and was canceled after a year.

The setting for *Twin Peaks* was the splendid Pacific Northwest landscape outside of Seattle, with the twin mountaintops really Mount Si. At a press briefing, at the launch of the show, Lynch said he was attracted to the area because of its abundance of Douglas Fir trees. Lynch said, "Just sort of picture this kind of darkness and this wind going through these needles of the Douglas firs, and you start getting a little bit of a mood coming along."

Initially, the producers had considered filming the show in Canada, and on *Twin Peaks* the location of the town is five miles south of the Canadian border, not in any particular state.

Several real locations in and around Snoqualmie, Washington, were featured on the show. The two-hour pilot was shot on location in the area, and crews came back regularly on a second-unit basis, shooting exteriors, mostly without the actors present.

## What to See in Snoqualmie, Washington

▶ **Salish Lodge,** 37807 Southeast Snoqualmie Falls Road. The exterior of this rustic hotel was used as the Great Northern Lodge on the show. Inside the real hotel are ninety-one luxury guest rooms. And the inn's five-star, award-winning restaurant offers three meals a day during the week and, on weekends, a special country breakfast until 3 P.M. and dinner service. Standard room rates start at $180 to $245, and suites start at $500 a night. The hotel has maps of *Twin Peaks* location sites available for guests. Call 800-826-6124.

The Salish Lodge was known as the Great Northern Lodge on *Twin Peaks*. *(Photo © George White Photography.)*

▶ **Weyerhaeuser Mill,** 7001396 Avenue Southeast. The office of this real mill appeared as the Twin Peaks sheriff station on the show. The exterior of the mill area is seen as the Packard Mill, owned by Widow Josie (Joan Chen), but wanted by greedy Catherine (Piper Laurie). The business is not open to the public, but fans can take pictures outside.

## What to See in North Bend, Washington

▶ **Mar-T Cafe,** 137 West North Bend Way. They really do serve cherry pie and "damn fine coffee" at this restaurant, which appeared as the Double R Diner, operated by Norma (Peggy Lipton), on the show. Filming for the pilot took place inside and outside the restaurant, which was later duplicated on the *Twin Peaks* set. Breakfast, lunch, and dinner are served with specialties including deli sandwiches and steaks. And, of course, pie is a big seller. The restaurant also sells *Twin Peaks* souvenirs including coffee cups, tins, T-shirts (which say "This Is Where Pies Go to Die") and postcards. Open seven days a week, 5:30 A.M. to 7:30 P.M., and until 8 P.M. on weekends. Call 206-888-1221.

▶ **Alpine Blossom and Gift Shoppe,** 213 North Bend Boulevard. This florist shop is the place to go for *Twin Peaks* souvenirs, which include posters, T-shirts, hats, buttons, magnets, tapes, special newspapers and magazines, postcards, coffee cups (which say "A Cup of Joe"), and even earrings (which depict a piece of pie and a cup of coffee). *Northern Exposure* items are also for sale (the town where that show was filmed is only about 60 miles away). Call 206-888-2900.

▶ **Snoqualmie Winery,** 1000 Winery Road. This local winery was the setting for the Laura video in the show's pilot. Eleven different wines, ten whites and one red, are produced here. The winery offers free tastings, and bottles are for sale. Open daily, 10 A.M. to 4:30 P.M. (closed Easter, Thanksgiving, and Christmas). Call 206-888-4000.

## What to See in Fall City, Washington

▶ **The Colonial Inn,** 4200 Fall City-Preston Road. This restaurant, which offers live musical entertainment on weekends, was seen as the Roadhouse on the show. Fine dining specialties include prime rib. The restaurant is open for three meals a day, seven days a week. Reservations are recommended on weekends. Open 7 A.M. to 8 P.M. on weekdays, and until 9:30 P.M. on weekends (lounge closes at 11 P.M.). Call 206-222-5191.

## What to See in Preston, Washington

▶ **Stonecroft Gallery,** 8606 Preston-Fall City Road, Southeast (just past the general store). This local artist gallery and antique shop was the setting for Big Ed's gas station on the show. In real life, a potter can be viewed in the shop making pots. The shop also sells *Twin Peaks* souvenirs. Open 10 A.M. to 6 P.M. on Monday, Thursday, Friday, and Saturday, and 1 P.M. to 6 P.M. on Sunday (on Tuesday and Wednesday, open by chance). Across the street is a private house that appeared as crazy one-eyed Nadine's (Wendy Robie) home on the show. Call 206-222-7687.

The *Twin Peaks* fan club holds an annual *Twin Peaks* festival in the Snoqualmie, Washington, area during the second week in August, which features appearances by the show's stars. For more information call 810-752-5142.

## Fun Facts

- A follow-up movie, *Twin Peaks: Fire Walk with Me*, attempted to solve some of the show's mysteries, but left a lot of areas vague.
- Cocreator David Lynch is a former Boy Scout who was born in Missoula, Montana.
- Cocreator Mark Frost said the show's small-town setting was partially inspired by the lakeside town in upstate New York where he spent summers as a child.
- Mark Frost told *People* magazine that Dale Cooper's appreciation of cherry pie comes from him, and "damn fine" coffee comes from David Lynch.
- The show was originally scheduled to be only eight parts (including the two-hour pilot episode).
- David Lynch's movie credits include *Eraserhead, Blue Velvet,* and *Wild at Heart.* Kyle MacLachlan, who played Dale Cooper, was also in *Blue Velvet.*
- Angelo Badalamenti did the music for *Twin Peaks,* which is mostly jazz. He also scored *Blue Velvet.*
- Peggy Lipton was earlier on *Mod Squad,* of which *Twin Peaks* cocreator Mark Frost was said to be a fan. Lipton, who, like several of the actors on *Twin Peaks,* had been in retirement for a few years before the show, said through the role of Norma she discovered her own capabilities as a survivor.
- David Lynch directed the pilot and a few episodes of *Twin Peaks,* and also appeared on the show in the role of FBI Bureau Chief Gordon Cole, who is hard of hearing.
- Michael Ontkean, who played Harry S. Truman, the sheriff of Twin Peaks, was earlier in *The Rookies.*
- Frank Silva, who played Bob, the personification of evil, was tapped for the part while a crew member on the show.

• *Twin Peaks* was a particularly big hit in Japan, where tens of thousands of sets of fourteen tapes of the show sold for $440 a pop, according to *U.S. News & World Report.*

• A local newspaper in the Snoqualmie area printed a letter written by a boy in Turkmenia, and addressed to the Mayor of Twin Peaks, that asks for information on Laura's (Sheryl Lee) personal life and hobbies, without the slightest hint that the boy realized the character was fictional.

# HAWAII

*Hawaii Five-O* may be the most popular show to feature Hawaii, but it's not the first. The lush tropical island landscape has attracted TV production crews since around 1959, when *Hawaiian Eye*, a detective show, went there to shoot exteriors, including shots of downtown Honolulu. The show's stars, Robert Conrad and Connie Stevens, stayed in Hollywood however, doing their work in a studio.

A few years later *Adventures in Paradise*, which starred Gardner McKay as a schooner captain, did the first full TV location shoot in Hawaii. On the show, however, the setting was supposed to be the South Pacific.

The comedy *Gilligan's Island* shot its pilot on location in Kauai, in 1963, and episodes of the show featured shots of a real island off Oahu, as well as a real Honolulu yacht club, in their intros. But most of the show's action really took place in a Hollywood studio.

*Hawaii Five-O*, the cop show starring Jack Lord, was the first TV series to make Hawaii its home base for production, and much of the island of Oahu appeared on the show. Up to twenty locations on the island were used in each episode, and interiors were done in a studio in Diamond Head, set up especially for the series.

*Hawaii Five-O* was followed closely by another Hawaii-based series, *Magnum, P.I.*, which used the same studio in Diamond Head, and also showed the scenery in and around Oahu.

*Jake and the Fatman*, the lawyer drama starring William Conrad, went to Hawaii in its second season (Jake left his job as a district attorney in California to work for the prosecutor's office in Honolulu), filming there for two years. The real Hawaii

Supreme Court building (at the corner of Punchbowl and Ala Moana Boulevard), was among real locations that appeared on the show.

*The Byrds of Paradise*, the short-lived 1993 ABC series starring Timothy Busfield *(thirtysomething)*, also shot on location in Diamond Head, as well as at other island settings.

*I Dream of Jeannie* was among the shows that visited the Hawaiian Islands for location footage.

## GILLIGAN'S ISLAND
**(September 1964–September 1967, CBS)**

"Now sit right back and you'll hear a tale . . ."

The tour on the S.S. *Minnow* was supposed to last only three hours "but the weather started getting rough" and . . . well, you know the rest of the story.

A shipwrecked gang consisting of a skipper, his first mate, and their passengers—a movie star, a high school teacher (the Professor), a millionaire and his wife, and a Kansas farm girl—are stranded for three long years on a deserted island. *Gilligan's Island* was not a hit with critics, and some of the stars even said they were embarrassed to be in it. But the show drew devoted fans, especially in reruns, and at one time was on the air in seventy-four countries.

None of the lead actors hit it big after *Gilligan's Island*, however, and some complained they were later typecast.

In his book *Here on Gilligan's Isle*, actor Russell Johnson, who played the Professor on the show and later appeared on *Dynasty* (and in Captain D's Seafood commercials), says he wishes he had never taken the part. Johnson also complained in his book that Jim Backus, who played millionaire Thurston Howell III told dirty jokes, and Tina Louise, the Lee Strasburg-trained singer/actress who played Ginger Grant, the movie star, was not pleasant to work with.

The deserted island on the show was supposed to be three hundred miles from Hawaii. A real Hawaiian island was seen on the show, but it was Coconut Island, which is not deserted and is located only a half mile offshore.

Most of *Gilligan's Island* was filmed on a sand-filled soundstage at CBS in Studio City, California. But the show's pilot, which did not air until 1992 (when it appeared on WTBS, the cable Superstation), was shot largely on location near Moloaa Bay in Kauai, in 1963. The location is not far from where scenes from the film *South Pacific* were shot. Part of at least two episodes of *Gilligan's Island* were also shot near Zuma Beach in Southern California.

The show was produced by Sherwood Schwartz, who, in 1969, launched another rerun classic, *The Brady Bunch.*

## What to See on (and off) Oahu, Hawaii

▶ **Coconut Island,** about a half mile off the shore of Oahu (on Kaneohe Bay). The island, which appeared in a distance shot as the deserted island on the show, is actually part private residential property and part research facility of the University of Hawaii. Access is limited to guests of the university or the island's owners, but visitors can see the island by boat. The bay and beach areas also appear on the show, according to a local tourism source.

## What to See in Honolulu, Hawaii

▶ **The Waikiki Yacht Club,** 1599 Ala Moana Boulevard. The ship set sail "from this tropic port" on Alawai Harbor, actually a private yacht club. The club is open mostly to members, but also offers reciprocal access to members of other yacht clubs. It also appeared on *Hawaii Five-O.* Call 808-949-7141.

## Fun Facts

- According to Russell Johnson's book, Raquel Welch was up for the part of Mary Ann, and Carroll O'Connor was considered for the role of Skipper. Jerry Van Dyke was a possible for Gilligan, and Dabney Coleman for the Professor.
- During the show's run, people actually wrote the Coast Guard asking them to rescue the stranded group.
- "The Ballad of Gilligan's Isle" was performed by the Wellingtons. It was not used in the pilot, for which the intro music was a hastily composed calypso song.

"The ship set sail . . ." from the tropic port of the Waikiki Yacht Club, on *Gilligan's Island*. *(Photo by Ray Pendleton.)*

• Several TV movies and two cartoon versions were produced based on *Gilligan's Island*. In *Rescue from Gilligan's Island*, the gang got off the island, but got stranded again on a reunion cruise.

• The first name of Gilligan (Bob Denver) was never revealed on the show.

• Actress Dawn Wells, who played Mary Ann, was a former Miss Nevada.

• *Gilligan's Island: The Musical* opened off-Broadway in 1992. It was coproduced and written by Sherwood Schwartz and his son, Lloyd.

• Bob Denver, who played Gilligan, and Alan Hale, Jr., who played the Skipper, followed *Gilligan's Island* by appearing together in the show *The Good Guys*, also on CBS. It was not as successful.

• Bob Denver had previously played beatnik Maynard Krebs on *The Many Loves of Dobie Gillis*.

• Playing the part of the Professor in the pilot was John Gabriel, who later played Seneca Beaulac on the ABC soap *Ryan's Hope*.

• The show never explained why the gang on the island had packed so many clothes for a "three-hour cruise." Mrs. Howell (Natalie Schafer), also known as Lovey, even packed her jewels.

• Jim Backus, who played Thurston Howell III, also did the voice of Mr. Magoo in cartoons.

• The Skipper's real name on the show was Jonas Grumby. The Professor's was Roy Hinkley.

## HAWAII FIVE-O
**(September 1968–April 1980, CBS)**

Steve McGarrett (Jack Lord) battled criminals and the underworld, and especially Wo Fat (Khigh Dhiegh), in this tropical paradise cop show, shot on location in Hawaii. McGarrett, who usually wore dark suits, not Hawaiian shirts, worked for a state police unit, reporting directly to the governor of Hawaii, Philip Grey (Richard Denning). He was based at the Iolani Palace, a real historic landmark building on Oahu. McGarrett's assistant in the pursuit of criminals was Danny Williams (James MacArthur), who McGarrett called "Danno." The show became so popular that "Book 'em, Danno!" became part of the American vernacular.

A trailblazer for location work, *Hawaii Five-O* helped make Hawaii a major tourist destination as well as a film production center, adding big bucks to the local economy.

The sunny setting didn't hurt the show's popularity either. *Hawaii Five-O* broke records as the longest-running cop show. In 1979, the show was seen by 300 million viewers in 83 countries each week.

The show shot all over the island of Oahu, as well as on other islands, with locations including Waikiki Beach, the Punchbowl, Diamond Head, the Ilikai Hotel, the University of Hawaii campus, Hanauma Bay, and Kahala Beach.

Much to the amusement of locals, chase scenes filmed on city streets did not always flow logically from point A to B. But they certainly looked real on TV. The producers set up a studio in Diamond Head, with the show's production office listed in phone books as *Hawaii Five-O*. This sometimes caused confusion among locals, with people calling thinking it was a real police unit.

The Hawaii Police Department cooperated with the show, and off-duty officers were employed as occasional extras and for crowd control. The police department also lent to the show (for a fee) real cars, uniforms, and weapons.

## What to See in Honolulu, Hawaii

▶ **Iolani Palace,** corner of King and Richards Street. Once the seat of Hawaii's state government, this ornate building, built in 1882, is the only structure in the United States that was used as a royal palace. Of Italian Renaissance design, the building was built for King Kalakaua. It is an historic landmark and is open as a museum. Among the attractions is the gilded throne room, with its original furnishings on display. Hours are Wednesday through Saturday, 9 A.M. to 2:15 P.M. Reservations are required. A fee is charged. Call 808-522-0832.

## Fun Facts

- Actor Khigh Dhiegh (pronounced KI-Dee), who played the criminal Wo Fat, is of Anglo-Egyptian-Sudanese ancestry, not

The Iolani Palace. *(Photo courtesy of the Hawaii Visitors Bureau.)*

Chinese, as is his character. But he did study I Ching, the oriental philosophy.

• In 1980 Wo Fat is finally sent to jail.

• Actor Jack Lord, who is from New York City, not only acted in the show but also wrote, directed, and produced, and he owned a piece of *Hawaii Five-O* too. In his spare time, Lord is a painter. His artwork has been exhibited in numerous galleries including the Whitney and the Metropolitan Museum of Art in New York.

• Andy Griffith *(The Andy Griffith Show)* and Carol Burnett *(The Carol Burnett Show)* guest-starred in 1972, as bad guys.

• Prior to his *Hawaii Five-O* role, Jack Lord appeared on Broadway in *Cat on a Hot Tin Roof.*

• James MacArthur, who played Danno, is the son of renowned actress Helen Hayes and playwright Charles MacArthur. Hayes guest-starred on *Hawaii Five-O.*

## MAGNUM, P.I.
**(December 1980–September 1988, CBS)**

Magnum (Tom Selleck), whose real name was Thomas Sullivan Magnum, was a former Navy man who served in Vietnam and then began working as a private eye. He lived on the Hawaiian island estate of wealthy mystery writer Robin Masters—Magnum provided security for secretive Masters in return for his rent. As part of the deal, fun-loving Magnum also got to toot around Oahu in Masters's $50,000 Ferrari.

The lush estate was run by stuffy Jonathan Quayle Higgins III (John Hillerman), a British military intelligence officer in World War II, who assisted Magnum in his investigations. (Higgins actually may have been Robin Masters).

*Magnum, P.I.* filmed on location in Hawaii, showing real tropical scenery and sites. Interiors were done on a soundstage in Diamond Head, the same one that was used for *Hawaii Five-O,* CBS's earlier hit Hawaii-based show (no coincidence here).

Steve McGarrett of *Hawaii Five-O* was mentioned on *Magnum, P.I.*, but actor Jack Lord, who played McGarrett, did not appear on the later show.

*Magnum, P.I.* sometimes left the island, shooting one episode on location, for instance, near Kent, England. The setting was historic Leeds Castle, which was built in 857 A.D.

## WHAT TO SEE IN HONOLULU, HAWAII

- **Kahala Mandarin Oriental, Hawaii (formerly Kahala Hilton Hawaii),** 5000 Kahala Avenue. What appeared as the fictional King Kamehameha Beach Club on the show is really a beachfront function area at this hotel. The area was built for the show, but filming here became a problem because the structure was too close to the hotel's parking garage, and there was background noise from cars and screeching tires. So filming took place at the hotel only for a year, a spokeswoman said. A plaque at the site indicates its TV history. The hotel is under renovation and is expected to reopen in 1996. Call 808-734-2211.

- **The private house** (off the Kalanianaole Highway, near Wiamanalo Beach). Public access is not permitted at this private residence on the Pacific Ocean, which was seen as the fictional Robin's Nest estate on the show. Sources said the house is the second house before the first beach entrance to Wiamanalo Beach. But it's tricky to find and not very visible from the street.

The Magnum Bar at the former Kahala Hilton (recently renamed Kahala Mandarin Oriental, Hawaii), known as the King Kamehameha Club on *Magnum, P.I.* *(Photo courtesy of the Kahala Hilton.)*

Said one local, "If the only reason for going to the area is to see the estate, you'll be disappointed. It's not even a very big house for Hawaii."

▶ **Tommy's Tours.** This local tour operator includes a quick drive by the entrance to the house that appears as the Robin's Nest estate, as part of its Beach Blast tour. The tour also features stops at less known beaches on the Island of Oahu. Call 808-944-8828.

## Fun Facts

• *Magnum, P.I.* made macho Tom Selleck a sex symbol and helped him find his way into a subsequent movie career. He had earlier appeared as a private investigator on *The Rockford Files.*

• The face of Robin Masters was never seen. The voice heard for several years as Masters' on the show was that of actor/director Orson Welles. In the final episode, Jonathan Higgins (John Hillerman) admitted he was Masters, but then took it back.

• Broadway star Gwen Verdon appeared as Katherine Peterson, Magnum's mom.

• Frank Sinatra guest-starred on the show in 1987, playing a retired New York City police sergeant.

• Jessica Fletcher (Angela Lansbury) of *Murder, She Wrote* visited Magnum in 1986; there was supposed to be professional jealousy between the two fictional detectives.

• *Simon & Simon* (Jameson Parker and Gerald McRaney) also visited *Magnum, P.I.*, in an episode which also guest-starred Morgan Fairchild as a thief.

• Magnum was shot and killed, and went to heaven, at the end of the 1987 season, but when the series was unexpectedly renewed he came back. As with the famous shower scene in *Dallas*, it was revealed that Magnum was only dreaming about going to heaven. In the real finale, in 1988, Magnum cut his hair short and rejoined the Navy.

• Tom Selleck, who won an Emmy for his Magnum role in 1984, also produced some episodes of the show. He later was executive producer for *B.L. Stryker,* which starred Burt Reynolds *(Evening Shade)* as a private eye.

# INDEX

(A page entry set in *italics* indicates a related photograph.)

Fran Wenograd Golden probably watched too much TV as a child. When she grew up she worked as a radio and TV reporter, wrote for publications including *The New York Times* and *Popular Mechanics* magazine, and, in 1984, joined the staff of *Travel Weekly*, the largest travel trade newspaper. She lives near Boston with her husband, two children, and a crazy dog, Lucy.